Railway Engineering

MAXWELL PRESS

Railway Engineering

Cecil B. Smith MA.E.,

Late Assistant Professor of Civil Engineering in McGill University,
Member Canadian Society of Civil Engineers.

MAXWELL PRESS

Chennai Trichy New Delhi

MAXWELL PRESS

This edition has been published in india by arrangement with Carson Books, UK

ISBN 978-93-5528-229-3

Maxwell Press
No. 44, Nallathambi Street,
Triplicane, Chennai 600 005

Printed and bound in India

MJP 1285

Publisher : C. Janarthanan

Publisher's Note

The legacy of a country is in its varied cultural heritage, historical literature, developments in the field of economy and science. The top nations in the world are competing in the field of science, economy and literature. This vast legacy has to be conserved and documented so that it can be bestowed to the future generation. The knowledge of this legacy is slowly getting perished in the present generation due to lack of documentation.

Keeping this in mind, the concern with retrospective acquiring of rare books has been accented recently by the burgeoning reprint industry. Maxwell Press is gratified to retrieve the rare collections with a view to bring back those books that were landmarks in their time.

In this effort, a series of rare books would be republished under the banner, "Maxwell Press". The books in the reprint series have been carefully selected for their contemporary usefulness as well as their historical importance within the intellectual. We reconstruct the book with slight enhancements made for better presentation, without affecting the contents of the original edition.

Most of the works selected for republishing covers a huge range of subjects, from history to anthropology. We believe this reprint edition will be a service to the numerous researchers and practitioners active in this fascinating field. We allow readers to experience the wonder of peering into a scholarly work of the highest order and seminal significance.

MAXWELL PRESS

This Book is respectfully dedicated

to

Sir William C. VanHorne,

President of the Canadian Pacific Railway Company,
who has so successfully directed its policy as to
make it a great National and Imperial
Highway, a blessing to the
country, and a finan-
cial success.

CONTENTS.

PART I.

LIST OF PLATES, DIAGRAMS, ETC.

LIST OF TABLES.

RAILWAY ENGINEERING.

INTRODUCTION.

This book is the outcome of an endeavor on the writer's part to epitomize a vast subject into such a compass that the student or layman whose experience is pre-supposed to be "nil" may grasp it in an intelligent way. It is intended to be a foundation course only, and as such, has been largely selected from the various works bearing on each department of the subject; but the proper balancing of the parts, if such there be, giving each its due importance, the combination of the whole subject, technically considered, as a ground work for future study, and the exclusion of much confusing detail which obscures the mental vision, the writer may claim as his own.

During the present period of depression, which always so seriously affects railway construction, it might be thought that the vocation of the railway engineer was being largely obliterated; but this is not at all a consequence. Our railways must be maintained, and while more engineers, per mile, are employed during construction than afterwards on maintenance, yet, although there are, no doubt, pleasant and remunerative positions to be had during the former period, there is no condition of permanence that makes them desirable. On the other hand, railway companies recognize more, every day, the value of a technical engineering training for those young men who fill junior positions in the operating and maintenance departments, not strictly engineering in their nature.

And those companies (*e.g.*, Pennsylvania, or Norfolk and Western) that have persistently filled such positions with young engineering graduates, that have had them do routine work and given them a business training, have seen their highest offices filled by men whose engineering knowledge has brought them to the front, when aided by a good business training, a knowledge of ways and means, and of traffic and operation.

In the future, such positions, and those on the maintenance staff proper, would seem to be the paths more likely to lead to success than the more strictly technical work of location and construction, particularly as the construction in future will be chiefly in the shape of short extensions of large systems having permanent staffs.

This book, however, will deal chiefly with location, construction and maintenance, not because these cover the whole ground, but because a knowledge of traffic, rates, operation and management can be gained only by experience, whereas a good grasp of the former may be had previous to employment of such an extent, at least, as will be valuable in obtaining and filling junior railway positions, and also form a basis for future study. And even though very little of what is here given may be used at once by the young engineer, yet it will enable him to take a more intelligent interest in all that his superior officer does, which he could not otherwise do unless he had a proper understanding of the general principles on which railways are surveyed, constructed and operated. He is warned against having his faith in these principles shaken by the adverse criticism of men who do not appreciate or understand them. Care is taken to give here only what is fairly well tried and established.

On the other hand, he is advised to keep his opinions, largely, to himself, and to carry out faithfully the instructions of his superiors in office. These instructions, though perhaps sometimes faulty, should be studied and respected, so that when the time arrives that he, in turn, gives orders which must be obeyed, he may put into practice what he then considers, after several years' experience, to be best, not only theoretically, but from the standpoint of being feasible and advantageous, capable of being put into execution by his assistants—the best, all things considered.

It must not be forgotten that this work is not exhaustive, but merely introductory. Years of reading, conversation, experience, observation, and above all, honest hard thinking, are necessary to complete a man's knowledge on any subject, and even then it is not complete. So that we must never desist, but always persevere, if we wish to keep up with the progress of this most progressive subject. C. B. S.

Montreal, Que., Canada, May, 1897.

CHAPTER I.

FUNDAMENTAL CONSIDERATIONS.

ARTICLE I.—TRANSPORTATION.

The inhabitants of the civilized world have, since the year 1825, been enabled to remodel their ideas of how and where to live. There has been developed within this period a new potentiality, which had through all the previous history of the world been practically dormant. The impulse given by it to the material and industrial progress of the world is such as to stamp it as one of the grandest events of our world's history, and it will be so spoken of in future ages. It is the development of transportation.

In its broadest sense, transportation may be said to include all means of communication; but of its various phases the transportation of material objects by means of the railway train will be the one treated of in this book.

The railway had its birth in England, and a fierce struggle took place between it and the canal for supremacy, while in North America the canal systems not being far advanced, and the extent of territory to be traversed rugged and vast, the result was never in doubt; to-day, the canal is a useful regulator of rates, and a means of transportation of heavy bulk freights in which time is not a factor, but it cannot be said to be a competitor of the railway to any serious extent.

By 1850 the people of North America had grasped the fact that the rapid extension of our railways to the remote and unsettled regions westward, was the key to that marvellous growth that has peopled a continent in so short a time. The capital available was small and the country fairly rough, so that different methods of construction and operation from those in vogue in England, and a consequent different class of equipment, were imperative.

At the present day, in Canada, our railways are developed along the same general lines as those of the United States, and in it we have done our fair share, but it must be recognized that to the civil engineers of the United States is due the credit of those essential departures from early forms which have defined our continental types so distinctly, and are the glory and the boast of North Americans. These departures took place gradually, the gap becoming wider every year, until now it has passed its maximum, and the slow conservatism of English engineers is yielding. Bogie trucks, equalizing levers, Westinghouse brakes, and American cars are becoming familiar in England, while on the other hand increased wealth and traffic are enabling American railways to introduce block-signalling and interlocking systems, to abolish many grade crossings, and make their road-beds more solid and permanent.

The distinctive features of the railway system of North America that have enabled it to extend to a length of over 200,000 miles (including Mexico and Central America), that have given Canada a system of over 16,000 miles, moving 22,000,000 tons of freight, 14,000,000 passengers, 60,000,000 newspapers, 100,000,000 letters, besides much express, etc., each year, having a capitalization of $900,000,000, and employing an army of perhaps 55,000 men, are as follows:

(1) A frank recognition of the fact that curvature is not a great drawback, and can be introduced freely to economize construction.

(2) The introduction of bogie and swivelling trucks and equalizing levers, enabling lines of poor surface and sharp curvature to be operated safely and economically.

(3) The use of long wheel-bases on engines for freight work, enabling greater weight to be put on the drivers of engines operating over quite inferior track.

(4) The consequent hauling of increasingly heavier loads of freight per engine and per train crew.

(5) The lowering of freight rates to a point that enabled coarse freights to be worth moving, thereby increasing the volume of freight enormously.

(6) The acceptance of a timber-construction period, enabling roads with meagre early traffic to pay their small fixed charges and survive until their finances and credit are such as to enable them from their earnings or by increased bonding to replace such structures with permanent ones. The Canadian Pacific Railway is a striking example of this.

(7) The use of increasingly heavier freight cars, in which paying freight is a larger percentage of the gross load—and also giving a less co-efficient of rolling friction —which the following table illustrates :

1875....20,000 lbs. car,	20,000 lbs. freight,	50 per cent.	dead load
1880....24,000 "	40,000 "	37 "	"
1890....28,000 "	60,000 "	31 "	"
1896....36,000 "	80,000 "	31 "	"

From which it appears that the limit has been reached.

ARTICLE 2.—PROJECTS.

A company of limited liability, but whose capital is inelastic and non-circulating, must do business or break down; it cannot contract its business in hard times except at a sacrifice; business at starvation wages is better than none, and this is the exact condition of a railway company which is a manufacturer and seller of transportation. In this it is different from a store, or more particularly a banking house, therefore all the more carefully should the project be studied before money is embarked in it.

No considerations of a general character will cover all cases, and therefore it will be necessary to exclude roads which have been or may be built (*a*) for purposes of blackmail, to force rival companies to buy them out ; (*b*) for speculation of the builders, not owners. These are not legitimate enterprises, but ones which projectors start by the expenditure of a small sum for charter, issue of bonds, etc., expecting to charge a margin for selling the bonds, to form construction companies, and let the contracts of construction to themselves at high prices, getting all the money out of the bondholders, running no risk themselves, but controlling all management by means of valueless stock. This gives them all the voting power, and any extra profit remaining after the bond coupons have been paid. Even such roads as these, however, will profit in the same way, as legitimate enterprises, by the application of true economy in location and construction.

Cost is the basis of all business, and most particularly in the case of railways must this always be so. An engineer may insist on technical accuracy and massive work, to such an extent as to bankrupt his company before the road is on a paying basis or even built, or he may, in an ill-directed effort toward economy, give it such a miserable constitution of grades and position, relatively, to its customers, that it will never secure traffic, and could not handle it economically if it did. Between these two extremes, the intelligent engineer should strike a happy balance, so that the project may be where it can obtain most traffic, at least first cost consistent with moderate working expenses, so that it will be profitable to the present owners or promoters, who usually build the road on borrowed money up to a certain safe mortgagable amount.

The promoters of roads are always sanguine, and probably the most common error into which such men usually fall, is to overrate the funds on hand or in view, and on the other hand, to underrate the cost of the completed enterprise. Roads are seldom built within their first estimated cost, and therefore, this is a danger against which the chief engineer must guard; he must be sure and firm in his figures, because it is difficult to foresee all contingencies, and still more so to impress the directors with the reality or necessity of each item.

The finances at the command of the company should always be fully known to the chief engineer; he has the right to know it, and should have the courage to insist on the fullest confidence of the directors. These means should be carefully studied, allowances made for changes in the money market affecting the value of bonds, the amount of money which can be raised easily, and the difficulty in getting the *last* part of the required amount should also be considered. Usually the bonds of new roads, just being built, sell below par, and, as the amount issued increases, the selling price may get less and less, until they may become unsaleable.

Many roads become bankrupt before or just after construction is finished, and a promising project ends in a receivership, the wiping out of past debts, or issue of prior-lien bonds on the part of the bondholders themselves. Receiverships, instituted originally to protect bondholders, are often made the instruments of defrauding them. The history of the railways of the United States, particularly, is full of examples of unnecessary roads built on faith and hope, and ending in disaster or fraud. Over 25 per cent. of United States railways are now in receivers' hands, and nearly all have passed through that stage in some period of their history.

The most casual observation teaches that in a country like Canada, where traffic is still unfortunately very light, we must build roads with the utmost economy. This has been practised in several justifiable directions.

(*a*) The introduction of curves where necessary, with a sharpness of as high as 4° to 6° on main lines, and 8° to 10° on branches, with a frequency only limited by a piece of tangent of 200 to 400 feet long between curves; in this way, by a slight addition to the cost of hauling trains and length of line, the cost of road-beds has been kept at a minimum.

(*b*) The use of fluctuating grades, by which the local "sags" or depressions do not increase the cost of hauling trains, but cheapen the cost of construction materially, and which have no objectionable feature except a change in train speeds, as they store up or yield a part of their "velocity head."

(*c*) Timber structures over all important streams, and even timber box culverts under light banks; in this way a railway company is enabled to get its road in operation quickly at a minimum cost, is able often to tide over the first few years of meagre traffic, replacing them, gradually, as means will permit, with permanent structures. On the other hand, there are certain directions in which economy cannot be practiced.

(*a*) Narrow gauge roads, except in isolated cases, have now been abandoned, because the demands for interchange of traffic put them at a disadvantage; because the cost of construction is higher in proportion to carrying capacity of cars, etc., and chiefly because it is found that American engines of standard gauge can pass around any ordinary curve quite freely.

(*b*) Light rails. This will be dealt with more fully in future chapters, but it may be well to say here that with rails quoted at $20 to $25 per ton, there is no greater blunder than to buy light rails. In stiffness, strength and wear the increase varies nearly as the square of the weight per yard, thereby decreasing maintenance charges enormously as the weight increases. The present weights are roughly 60 lbs. per yard for branches and 80 lbs. for our main lines, with a strong tendency upwards.

(*c*) Excessive ruling gradients. Almost any other mistake can be corrected in time, curves can be flattened, short grades lifted, temporary structures replaced, but the ruling grade is the life or death of a road that has or expects to have any traffic beyond a meagre minimum. This question will be fully dealt with in Chapter II.

(*d*) Locating roads adjacent to but not through towns. Many instances might be given of this fact, where railway companies, in order to save money on right of way, to shorten the line slightly, or out of pique at not receiving bonuses, have built the road a mile or more away from the centre of population. Experience proves, however, that it is usually profitable to pass as near as possible through the *very heart* of all towns or cities, even at considerable extra expense.

The engineer must, therefore, when entrusted with a study of proposed routes, have several leading ideas constantly in his mind:

(1) How to obtain the most traffic, including the idea of shutting out, avoiding or fighting competitors.

(2) How to get a road built with as small fixed charges as possible consistent with small operating expenses, and clause (1).

(3) How to build a road that will be operated and maintained at as small a charge as is consistent with clauses (1) and (2). *These three things are intimately intertwined*, but may be affected by such considerations as obtaining heavy local aid, having heavier grades in direction of lesser traffic, and a complete change of train loads at the end of each engine division (100 to 130 miles), excepting always that the whole road will allow the passage of moderately heavy passenger trains intact.

Unfortunately these matters are often, erroneously enough, may be, settled quite apart from engineering ideas, politics and local aid being the controlling factors; but facts remain, and while politicians perish and local aid, once given, looks for a *quid pro quo*, the railway burdened with too heavy grades, too much debt, or distant from its customers, will gradually, but surely, fail in the race. The problem which has to be solved, in each case, is to create a paying property without satisfying, often, the dangerous desire on the part of the engineer to build solidly and erect monuments to himself, or satisfy his innate desire for excellence of construction considered from too narrow a standpoint. This is a difficult matter in a thinly settled country like Canada, as statistics to be given will show, but our roads are being more economically constructed and operated day by day and traffic is slowly increasing, so that we may confidently look forward to a time when there will be a change and some small returns for the stockholders and promoters.

ARTICLE 3.—TRAFFIC.

Wellington demonstrates that the traffic revenue increases with the (*population per mile of railway*) [2].

This is based on the rough assumption that the volume of traffic increases as the distance between two towns diminishes, or that the gross traffic receipts between two towns is nearly a constant, and thus if on a given line we have two traffic points and call traffic

1. Then with three traffic points the traffic = 1 + 2, and with n traffic points the traffic $= n\frac{(n-1)}{2}$, or when n is large we may neglect the second term and say that the traffic for n points $=\frac{n^2}{2}$. Now if we apply this to the individual as a unit, we may deduce the general statement given above. This assumption is not tenable when applied to a special commodity which originates at a fixed place, such as coal. Because the traffic is the same for two towns 150 miles or 15 miles from the coal pit, depending entirely on the demand for coal, on the other hand it is augmented by the fact that short haul rates are usually higher than long haul for the same service. Even this consideration, however, will not hold at the present day for suburban steam traffic, because it is being terribly crippled by electric suburban railways. On the whole, it is probably still true for a road of considerable length of general traffic, not largely suburban.

This view is upheld by the following table of the internal traffic of New York city:

Year.	Population.	Trips per year per inhabitant	Value by $(n)^2$ Formula
1860	814,000	45	45
1870	942,000	122	60
1880	1,206,000	175	99
1885	1,393,000	213	132

By which we see that the gross returns exceeded the ratio of $(population)^2$, but as the length of haul also increased it is probable that the net revenue about followed the law given. The following table, also, of a broader and more general character, confirms the view given :—

TABLE I.

SHOWING INCREASE OF POPULATION AND RAILWAY EARNINGS IN UNITED STATES AND CANADA, 1870 TO 1895

	Year.	Miles of Railway.	Gross Earnings ÷ 10,000	Population ÷ 10,000	$(\text{Col. iv.})^2$	Rates of Col. iii. Col. v.
Canada.	1875	4,300	$1,958	39.	1,521	1.29
	1880	6,800	2 355	43.	1,849	1.27
	1885	10,200	3 223	45.	2,025	1.59
	1890	13,100	4,680	48.	2,304	2.03
	1895	16,091	4,680	51.	2,601	1 79

	Year.	Miles of Railway.	Gross Earnings ÷ 10,000	Population ÷ 10,000	$(\text{Col. iv.})^2$	Rates of Col. iii. Col. v.
U.S A	1870	54,000	37,000	385	148,225	.249
	1875	74,000	50,000	440	193,600	.258
	1880	94,000	60,000	501	251,000	.239
	1885	128,000	77,000	565	319,225	.241
	1890	166,000	120,000	626	391,876	.307
	1895	180,000	107,500	690	476,100	225

This table shows (assuming that the population served per mile of railway increases in the same proportion as the population of the country as a whole does) that the gross earnings, both in Canada and the United States, have increased more rapidly than the $(\textit{population})^2$; however, the rapid falling off from 1890 to 1895 in both countries suggests that this may not entirely hold true in the future.

As another indication of the same law, compare Canada and United States for 1895. (See Table II.)

	Gross Earnings per Mile.	$(\text{Pop. per mile of Railway})^2$
Canada..................	$2,908	$(317)^2 = 100,489$
U.S.A.	5,945	$(382)^2 = 145,924$

Which shows that the gross returns per mile of railway are even more than in the proportion of "square of population per mile of railway."

In any given case it will be difficult to estimate the actual tributary population, which may be decreased by competition, increased by feeders or affected by industrial conditions, but if applied on a large scale to eliminate irregularities, we may look on it as quite accurate enough to guide us in comparisons of routes. Thus, supposing a railway by altering its general location can be made to serve 1,200,000 people instead of 1,000 000, we may estimate its gross returns to be increased from 100 to 144, or 44 per cent., while if the road is not materially longer, or poorer physically, the increase in operating expenses will be much less in proportion.

Revenue is often considered to be injured by a long line between terminals, but this is, largely, a mistaken idea ; the only case in which it is true is when there is keen competition between two places, by two or more routes, in which case the rate is fixed by the shorter line, but in non-competitive and local traffic, or even in the case of a road forming one link in a trunk line, rates are fixed by the mile or divided on a mileage basis, so that if a road can be located through more populous districts, built cheaper or with lighter grades, then the advantages of higher traffic charges, greater volume of traffic, less working expenses and fixed charges may any or all be in favor of the longer line. Incidentally, in Canada bonuses are often given on a mileage basis. These remarks apply only to moderate increases in length of line of from 5 to 10 per cent.

The folly of long tangents, creating more first cost, and often missing local traffic points, is a blunder somewhat common; the idea of serving the public by passing through the *very heart* of each populous district should be more fully appreciated, for ultimately, whatever traffic a road may have is from door to door, including cartage and bus fares; most evidently is this so at competing points. But the most important effect is that a good railway service convenient to the public will foster and increase traffic, while a town given the go-by for the sake of saving a little in land damages, or distance, has often had its prospects blighted forever; it is most important that a road should establish large roomy depots and obtain plenty of yard room while land is inexpensive, in anticipation of future growth on the part of any prosperous town. Wellington estimates the loss of traffic for each mile a depot is distant from the centre of population at from 10 per cent. to 25 per cent., being greatest at competing points and in manufacturing towns.

TRUNK LINES.

Most trunk lines are liable to suffer from competition, and, to protect themselves, buy or build feeders, and this has concentrated railways into large systems, but a general rule is to link together the largest possible population, quite regardless of minor losses in distance. The limit should never be approached when the increase in revenue is no greater than the increase in length of line, or when differences of distance are so great as to discourage traffic or encourage the construction of a competing road, and an exception might also be made in the case of being able to pass midway between two towns and serve them fairly well by branches.

The N.Y.C. & H.R.R. is a striking example of a road much longer than any of its competitors, but with light grades and a heavy tributary population, it is largely independent of its through traffic, and can handle it as "excess traffic" at a very low rate; on the other hand, the Pennsylvania R.R., soon after its completion west, built feeders in every direction, and thus held traffic that would otherwise soon have passed into other hands owing to its heavy grades.

Other general conclusions regarding trunk lines are:

(1) That they should never attempt to make a small sea or lake port a terminal, but have the largest possible terminals even at the expense of considerable extra distance. As instances of this, witness the Intercolonial making arrangements to enter Montreal, the Erie Railway abandoning Dunkirk as a terminal and building into Buffalo, and the Mexican National attempting to establish a port at Corpus Christi, instead of Galveston, which proved a failure.

(2) That after joining together as large a population as possible without unduly lengthening their line, they should build or buy such a system of branches, as feeders, as will draw to the main line as large a volume of traffic as possible, even in the face of competition. No better example of this can be given than in the Province of Ontario, where the G.T.R. and C.P.R. both endeavor to have feeders in all directions, and fight each other in many towns. Almost all the independent small lines of that Province have disappeared.

Branch Lines.—Branch lines are usually unprofitable in themselves, *e.g.*, Midland Railway of Canada, a network of short lines operated by the G.T.R. at a yearly loss, for the purpose of securing a large volume of freight for its main line, and preventing the C.P.R. from getting it. This is the reason, in almost all cases, which causes the most prosperous roads to operate them, to swell the trunk line traffic, because once any

TABLE II.—RAILWAY STATISTICS, 1895.

CAPITALIZATION, FIXED CHARGES, WORKING EXPENSES, EARNINGS, ETC.

Item.	Canada.		Great Britain.		United States.	
	Total.	Per Mile.	Total.	Per Mile.	Total.	Per Mile.
Mileage	16,091	....	21,174	....	180,657	§
Population	5,100,000	317	39,300,000	1,856	69,000,000	¶382
Square miles	1,609,000	....	120,800	....	2,970,000	¶
Population per square mile	3 2	....	325	....	23.2	....
Capitalization.						
Bonds	$330,786,000	$20,557	$1,800,906,000	$85 000	$5,407,000,000	$30,000
Preferred stock	105,680,000	6,568	1,236,446,000	58,400	759,000,000	4,200
Common stock	255,769,000	15,897	1,772,862,000	83,700	4,961,000,000	27,400
Loans and floating debts	37,626,000	2,338	67,196,000	3,100	616,000,000	3,500
Bonus and Government aid	167,523,000	10,411	....	....	....	....
Total	$897,384,000	$55,761	$4,875,410,000	$230,200	$11 743,000,000	‖$65,100
Earnings, Etc.						
Passenger	13,311,440	827	181,949,000	8,593	252,000,000	1,390
Freight	29,545,490	1,836	214,450,000	10,033	730,000,000	4,040
Other	3,928,560	245	22,044,000	1,136	93,000,000	515
Total	$46,785,490	$2,908	$418,443,000	$19,762	$1,075,000,000	$5,945
Working expenses	32,749,670	2,035	233,159,000	11,011	726,000,000	4,020
Net earnings	$14,035,820	$873	$185,284,000	$8,751	$349,000,000	$1,925
Per cent working expenses to gross earnings	.70	....	55¾	....	67½	....
Per cent. net earnings to total capitalization	1.56	....	3.80	....	2 97	....
Fixed Charges.						
Bond interest	*$15,287,250	(4 62%)	‡$121,494,000	(4%)	$240,000,000	(4.42%)
Other interest on loans, rentals, etc.	†1,881,300	(5 0%)	‡ 2 688,000	(4%)	53 000,000	....
Total	$17,168,550	....	$124,182,000	....	$293,000,000	....
Net income	— 3,132,730	....	61,104,000	....	56,000,000	....

TABLE III.—RAILWAY STATISTICS, 1895.

TRAFFIC, EQUIPMENT, ETC.

Item.	Canada.		Great Britain.		United States.	
	Total.	Per Mile.	Total.	Per Mile.	Total.	Per Mile.
Gross earnings per inhabitant	$9 00	....	$10 60	....	$15 60	
Passenger Traffic.						
Number of passengers	13,987,580	870	930,967,000	43,900	543,000,000	3,066
Passenger cars	1,994	$\frac{12}{100}$	42,230	2.	33,112	$\frac{18}{100}$
Average haul (miles)	38*	....	8	....	23	
Average load	35*	....	49	....	39	
Passenger cars per 1,000,000 passengers	143	....	45	....	65.	
Passenger train miles	15,332,276	....	184,200,000	....	327,294,000	
Passenger trains per day (each way)†	$1\frac{1}{4}$	....	12	....	$2\frac{1}{2}$	
Freight Traffic.						
Tons of freight	21,524,421	1,340	334,230,000	15,800	763,000,000	4,230.
Freight cars	52,118	....	603,710	....	1,196,119	
Average haul (miles)	110*	....	35	....	116.	
Average load (tons)	119*	....	73	....	180.	
Freight cars per 1,000,000 tons	2,421	....	1,806	....	1,717	
Freight train miles	19,939,699	....	150,400,000	....	491,410,000	
Freight trains per day (each way)†	$1\frac{7}{10}$	....	$9\frac{7}{10}$	....	$3\frac{7}{10}$	
Mixed Traffic.						
Various kinds of cars ‖	5,999	....	31,088	....	41,330	
Mixed train mileage	5,389,915	....	4,300,000	....	15,457,000	
Mixed trains per day (each way)†	$\frac{1}{2}$	....	$\frac{8}{10}$	....	$\frac{1}{8}$	
Total trains per day (each way)†	$3\frac{46}{100}$	....	22	....	$6\frac{83}{100}$	
Engines.						
Total number	2,023	......	18,658	....	35,111	Including switching engines.
Average yearly engine mileage‡	20,100		18,160	....	23,760	

branch line traffic is delivered to the main line, it being *extra* traffic, is very profitable, often costing almost zero to handle, while the toll collected is for the whole trip. Putting the cost of handling the unit of minimum traffic on a main line at 100, the cost of handling a single extra passenger or small parcel of freight is almost zero: in car load lots at 10 to 30 per unit, and in train load lots at about 50 per unit.

A branch line, however, usually costs nearly as much per mile to build as a main line, and nearly as much to maintain—certainly far more than in proportion to the traffic; so that we can usually lay down a rule of branch line location to make it strike the main line as soon as possible, but meet it at a town if possible in order to give the branch more return freight; country junctions of branches are deadly in their lack of traffic (other things being equal), therefore branch lines should rather be numerous and at right angles, or nearly so, to the main line, than to run parallel to the main line for any great distance, stringing together unimportant hamlets before joining the main stem.

Volume of Traffic.—The volume of traffic by Table III. is shown to be about $9 per head, per year, in Canada, and $15.60 per head, per year, for the United States. We may not assume, however, that it is uniform in Canada; but varies probably as rapidly as the square of the density of tributary population. The average town would probably be about $10 per year, per head, increasing to very much more for large towns and cities. The class of town also has a great effect; an industrial town such as Galt, Ontario, would afford far more traffic per head than such a town as Whitby or Cobourg, owing to the pursuits of the inhabitants being different. The great volume of suburban traffic cannot be counted on in the future, as the cheap roadbed, etc., few restrictions, frequent service, and convenient depositing of passengers, enables electric lines to serve such a traffic very successfully. This is a serious problem for steam roads to face, as suburban traffic has been, in the past, a very profitable feature to many roads.

INCREASE OF TRAFFIC.

The per cent. which operating expenses bears to the gross revenue varies enormously. Roughly speaking, the road which economizes in its investment, on which it pays interest, is apt to have heavy operating expenses. This percentage in Canada varies from 256 per cent. to 43 per cent., with an average of 70 per cent. for Canada in 1895.

These working expenses may be roughly divided as follows:—

(1)	Maintenance of line and buildings	15	70
(2)	Working and repairs of engines	22	
(3)	" " cars	6	
(4)	General expenses	27	

Now careful estimates show that only about half of these expenses are increased by an increase of traffic beyond a meagre minimum, which is the reason why it is so important to select a route giving the most traffic, as it is the *increase* of traffic over that which gives profit enough to pay fixed charges, to which we must look for profit to the stockholders, and a very moderate difference in first cost, revenue or working expenses means success or failure.

In any young country like Canada traffic increases rapidly at first, the increase being twofold: (1) A natural increase due to increased population; (2) An increase fostered by the newly discovered wants of a people not before served by a railway, the critical period of a road's history being usually the first few years of its existence, before a solid, steady revenue has been secured.

In England, New England States or Eastern Canada, the growth of traffic may be estimated at 5 to 6 per cent. per year for a given line. While in Western America or any new country, 10 to 15 per cent. per year will not be too much to figure on. The usual way of estimating traffic is by the number of trains per day over roads of certain maximum grades; but on roads of small traffic, which do not wish to run less than one train per day each of freight and passengers, the trains are not apt to be loaded well, and again two trains per day will be considered necessary to accommodate the people long before they will be regularly filled, so that it is only on roads of heavy traffic that it can be divided into the number of trains that will just accommodate it.

ARTICLE 4.—RECEIPTS.

Referring to Table II., "Railway Statistics," it will be seen that American freight charges are lower than English with a less volume of traffic. This anomaly is explained by the frequent, lightly-loaded freight trains of England run at a high speed, with small cars and heavy terminal charges. An adoption of heavy American cars, larger trainloads, and a slightly decreased speed, with perhaps five freight trains per day instead of ten, would enable English freight rates to be lowered more than one-half, and effect an enormous economy. Partly due to high rates on freight, but chiefly due to enormous traffic, the receipts on English roads are nearly $20,000 per mile per year, as against $5,900 per mile per year in the U.S.A, and $2,908 per year per mile in Canada; operations for 1895 showed net earnings to be 3.8 per cent. interest on gross capitalization of English roads, as compared with 2.97 per cent. per year in the U.S.A., and 1.56 per cent. per year in Canada.

This great difference, in spite of the capitalizations of the railways, per mile of railway, being $230,000 for Great Britain, $65,000 for U.S.A., and $55,760 for Canada, is due to two distinct causes, (1) volume of traffic, (2) decreased percentage of operating expenses to gross earnings due to this increased traffic, the net earnings being $8,751 per mile for England, $1,925 for U.S.A., and $873 for Canada. Let us now discuss the position which Canadian railways occupy financially, first taking Canadian railways as a whole, and, second, classifying them :

ARTICLE 5.—CANADIAN RAILWAYS.

The returns for 1895 are about as poor as could be selected, and those just handed down for 1896 show a slight improvement, but this may be due to temporary retrenchments, which have to be made up for sooner or later, such as track and car economies, by letting the condition run down slightly. The net earnings of Canadian railways of $873 per mile give net receipts of $14,035,820, but the interest on bonded debts, and estimated amount of loans, etc., amounts to $17,168,000* (approx.), or a net loss of over $3,130,000, and nothing with which to pay interest on the $167,000,000 that the various Governments have given either in the form of an investment in Government railways or as bonuses, and nothing with which to pay dividends on stock. This does not look encouraging; competition and popular clamor keep rates down to the lowest possible notch, and so long as our traffic is small it seems certain that, considering the severe climatic conditions under which our roads are operated, and the evident economy of management (70 per cent. of gross receipts as compared with 67½ for U.S.A.), the present rates must be fully maintained, if not raised. The stock of Canadian railways is not all real, but estimating that one-half is real and one-half water, we have $180,000,000 invested bearing no interest, besides the $167,000,000 which the people of the country have sunk in them in order to have sufficient railway accommodation. Even neglecting interest on all loans, floating debt, etc., there is still a deficit on bond interest of $1,252,000, which is met year by year by issues of more bonds and stock, or by creating floating debts which are periodically so converted; we go on year by year mortgaging futurity. It is hoped that in a few more years increased traffic will enable the bond interest to be fully met, and in this the increased solidity of permanent way will greatly aid. The average cost per train mile is very low (80½c.), considering our high price of coal, severe climate, rather inferior road beds, small number of trains per day, and good wages paid; it reflects great credit on Canadian management as a whole.

Let us now analyze Canadian workings by dividing the roads into four groups. (See Table IV).

* This is based on the assumption in Table II. that the interest on loans, floating debts, etc., is 5 per cent.; probably a large proportion of it bears no interest.

TABLE IV.—ANALYSIS OF WORKINGS OF CANADIAN RAILWAYS FOR 1895.

	CANADIAN PACIFIC RAILWAY SYSTEM.					
	Main Line.		Branches.		Total.	
	Total.	Per mile.	Total	Per mile.	Total.	Per mile.
Miles	3,575		2,709		6,284	
Capitalization—						
Bonds	$98,923,000	$27,700	$47,027,000	$17,300	$145,950,000	$23,200
Preferred stock	6,424,000	1,800	3,585,000	1,300	10,009,000	1,600
Common stock	65,000,000	18,200	12,410,000	4,600	77,410,000	12,300
Loans and floating, etc	1,948,000	500	3,445,000	1,300	5,393,000	900
Bonus and Government aid	58,996,000	16,500	17,454,000	6,400	76,451,000	12,200
Total	$231,291,000	$64,700	$83,921,000	$30,900	$315,213,000	$50,200
Interest on bonds	4,122,000		2,285,000		6,407,000	
Interest on loans and floating, etc	97,000		173,000		270,000	
Apparent total fixed charge	$4,219,000	$1,180	$2,458,000	$900	$ 6,677,000	$1,060
Gross earnings					17,962,000	2,860
Working expenses					11,317,000	1,800
Net earnings					+$6,645,000	$1,060
Per cent. of working expenses to gross earnings					63	
Apparent net income					— 32,000	
Train mileage					12,319,500	
Average trains per day					2⅔	
Earnings per train mile					$1 45	
Cost of train mile					91	
Operating expenses—						
Maintenance					15⅓ } 63	
Working and repair of engines					19	
Working and repair of cars					4	
General expenses					24⅔	

GRAND TRUNK RAILWAY SYSTEM.						Dominion Government Railways.		Remaining Railways of Canada.	
Main Line		Branches.		Total.					
Total.	Per mile.	Total.	Per mile.	Total.	Per mile.	Total.	Per mile.	Total.	Per mile.
884		2,280		3,164		1,382		5,261	
$ 75,243,000	$ 85,100	$33,649,000	$14,800	$108,892,000	$34,400			$75,944,000	$14,400
89,244,000	100,900	2,556,000	1,100	91,800,000	29,000			3,871,000	700
109,383,000	123,700	50,000		109,433,000	34,600			68,926,000	13,100
15,143,000	17,100	76,000		15,219,000	4,800			17,014,000	3,200
		10,301,000	4,500	10,301,000	3,300	$58,759,000	$42,500	22,012,000	4,·60
$289,013,000	$326,800	$46,632,000	$20,400	$335,645 000	$106,100	$58,759,000	$42,500	$187,767,000	$35,600
3,385,000		1,535,000		4,921,000				3,959,000	
757,000		4,000		761,000				851,000	
$4,143,000	$4,690	$1,539,000	$680	$ 5,682,000	$1,800			$4,810,000	$910
				16,091,000	5,090	$3,090,000	$ 2,240	9,642,000	1,830
				11,094,000	3,510	3,170,000	2,290	7,168,000	1,370
				+$4,997,000	$1,580	— $ 80,000		+$2,474,000	$460
				69		103		74	
				−685,000		—80,000		—2,336,000	
				15,381,200		4,230,000		8,732,000	
				6⅔		4 1-5		2¼	
				$1 05		73c.		$1 10	
				72c.		75c.		82c.	
				11½ } 69		25 } 103			
				24		36			
				7		14			
				26½		28			

(1) *Canadian Pacific Railway System.*—This road shows a low bonded debt, moderate stock, and high Government aid per mile (much of this latter, however, being tied up yet as land grant); on this account, and because it can charge higher rates in the west, it is enabled to keep its net earnings large enough to pay bond interest, and on prosperous years has considerable left over to pay dividends on stock. If dividends were declared in 1895, they were drawn quite justifiably from the surplus; but on the whole, the C.P. Railway by a very efficient management, high train-mile earnings ($1.45), and low first cost, is by far the most prosperous of Canadian railways, in spite of such low gross earnings as less than $3,000 per mile. The future increase of traffic in the North-West will enable it to lower its rates and still make a profit; but for the present, it is hardly clear how this can be done. The C. P. Railway also has the advantage of new equipment which does not need such heavy repairs. It is gradually improving its road-bed by the construction of masonry and steel in place of wooden structures.

It is noticeable that the cost per mile of the main line, $64,000, is not as high as the average of U.S.A. railways, and the route is very rugged, while the branches containing many first-class roads in Ontario and elsewhere have only cost $30,900 per mile, and very little, if any, has been charged against the main line, as in the Grand Trunk system.

(2) *Grand Trunk Railway System.*—The main line, built in early days when wages and material were high, built too solidly for its traffic, built in disregard of what is now known to be the true principles of railway location, and since then, burdened still further to pay bond interest and buy branches, is loaded with an enormous debt. The branches were often injudiciously bought, or bought to protect its traffic at a sacrifice, and do not pay. Taking the road as a whole, it is loaded with a heavy debt; suffers from keen competition, not only from Canadian, but U. S. railways; and by being thus forced to haul at low rates, has never been able to meet its obligations in spite of a traffic of

$5,090 per mile, which increases to $6,000 per mile in good years. Its operating expenses are higher, and train earnings lower than they should be, and a more vigorous management with a more careful study of maximum train loads by the tonnage system may effect improvement. This system has been the backbone of Canada; it has received very meagre aid ($3,300 per mile) from governments or municipalities, and the bondholders and owners deserve the highest praise for the integrity of management which has characterized it; a less scrupulous one would have thrown it into a receiver's hands after having plundered it for a few years. This has been done in America very often, with a severe loss to British bondholders. It is noticeable that the branches have been bought, largely, by mortgaging the main line; it is impossible to arrive at the cost of each.

(3) *Dominion Government Railways.*—These cost as much to operate as the earnings amount to. The cost per mile is a good example of actual cost, there being no inflation of capital. The Intercolonial was well built through a rough country; but could be built to-day for less than $30,000 per mile, owing to less cost of explosives, steel, labor, etc. It seems evident that for the light traffic of $2,240 per mile, there are too many trains per day; 4½ trains per day for 365 days is too much, and although the people on the route might object, the economy of reducing traffic to about three trains per day, say one passenger, one mixed, one freight would effect a great saving; this is evident by the very small train-mile earning, 72c.

(4) *Remaining Railways of Canada.*—These contain several high class roads, such as Canada Atlantic and Michigan Central; but on the whole, they are those roads which have been built to serve local needs or to develop country; their cost per mile ($35,500 being total capital against them) is small, but traffic still smaller; their earnings are not nearly sufficient to pay fixed charges, and some borrow more each year to pay it, while others default payment. The future of many of them is not reassuring until their traffic in-

creases ; but they are, on the other hand, many of them aiding in the development of country that would otherwise be beyond reach.

The need for more railways in Canada, except for some very special reason, is not at present apparent.

CHAPTER II.

TRAIN RESISTANCES AND THEIR COST.

ARTICLE 6.—TRAIN RESISTANCES.

It is necessary in a study of routes, especially in instituting comparisons of the advantages of alternative ones, to understand the nature, amount, and cost of the resistances which are offered to the hauling of trains, inasmuch as they have a direct effect on the working expenses of a railway.

These resistances may be any or all of the following, depending on circumstances; but the first four always exist in operating over even a level straight track.

A.—(1) Journal friction in the car trucks.
(2) Rolling friction, on the rails, of the car and engine wheels.
(3) Incidental—stopping and starting resistances.
(4) Velocity (wind and oscillating) resistances.

B. Grade resistances.

C. Curve resistances.

(*A.*) *Level Tangent Resistances.*—In the earlier years of railways, due to imperfect track, workmanship, etc., the journal and rolling friction was found by Clark to be about seven pounds per ton (2,000 lbs.); since then, however, the researches of Wellington and others have shown that on a good track, at ordinary speeds of over 10 miles per hour, these resistances are :—

Loaded Cars—4 lbs. per ton in summer.
" —6 " " winter.
Empty Cars—6 " " summer.
" —8 " " winter.

A change of 60° F. in temperature, showing an increase of 50 per cent. due to poor track, ice, snow, etc., and to an inherent increase in the co-efficient. Morin's law, that friction is independent of pressure or velocity, was deduced from data of a limited range, and does not hold true for extreme cases, the empty cars showing a decidedly higher co-efficient ; but when we consider the co-efficient in its relation to velocity the matter becomes more important. At the point of starting a train, the co-efficient is found to be as high as 18 to 20 pounds per ton, for loaded cars, and even higher for empties, being composed of friction + stiction ; the latter disappears the instant that motion begins, and the co-efficient falls suddenly, to perhaps 10 or 12 pounds per ton, at one or two miles per hour, and goes on steadily decreasing as the speed increases ; but above 10 miles per hour the change is not very appreciable. Should a train get "stalled" on a grade, or be at rest at a depot, it is plain that the *maximum* load cannot be hauled on the *maximum* grade ; this matter will be taken up further on.

It will thus be seen that under ordinary cases the force necessary to propel one ton along a level, straight track at ordinary speeds, and disregarding velocity resistances, varies from 4 to 8 lbs., depending on the time of year, condition of load, track and rolling stock ; in future calculations in these papers it will be taken at 6 lbs. per ton for loaded train.

The resistance offered to movement through space by the air has been extensively experimented on. The weight of evidence until recently was in favor of the belief that the resistance varied as $(\text{velocity})^2$. Based on this belief, the formula deduced for total level, tangent resistance on railways, by Clark, was :—

$$\text{Lbs. per ton hauled} = R = 7 + .0052\,V^2 \qquad (1)$$

(V being in miles per hour), in which the latter term represents the effect of the wind. Wellington also, as a result of the Burlington tests, made at moderate speeds, gave the following empyrical formulæ :—

$$\text{Lbs. per ton (a) for loaded flat cars} = R = 4 + .0065\,V^2 + .57\frac{V^2}{W} \qquad (2)$$

$$\text{" \quad " \quad (b) " box or psgr. cars} = R = 4 + .0075\,V^2 + .64\frac{V^2}{W}$$

$$\text{" \quad " \quad (c) " empty flat cars} = R = 6 + .0083\,V^2 + .57\frac{V^2}{W}$$

$$\text{" \quad " \quad (d) " box or psgr. cars} = R = 6 + .0106\,V^2 + .64\frac{V^2}{W}$$

Where the 2nd term represents the head and side air resistances, and the 3rd term the effect of oscillations, which being chiefly in the engine, decrease, per ton, as the train gets longer, V = miles per hour, W = gross weight of train in tons.

TABLE V.

POUNDS PER TON FOR TOTAL LEVEL TANGENT RESISTANCE

Speed in Miles per hour.	Clark (1) $7 + .0052 V^2$	Wellington 2(b) $4 + .0075 V^2 + .64 \frac{V^2}{W}$ (W = 450 tons.)	Wellington (3) $2 + \frac{V}{4}$	Average
10	7.52	4.89	4.5	6
20	9.08	7.56	7.0	8
30	11.68	12.03	9.5	11
40	15.32	18.27	12.0	15
50	20.00	26.30	14.5	20
60	25.72	36.12	17.0	26
70	32.48	47.71	19.5	33
80	40.28	61.08	22.0	41
90	49.12	76.27	24.5	50
100	59.00	93.20	27.0	60

Recently, however, the very high speeds obtained on various trial runs, and even scheduled trains, have thrown doubt on the idea that at high speeds, at least, the resistances could vary as the (Velocity)^2. In 1892, Mr. Wellington, after a study of Dudley's experiments on the New York Central Railway at speeds of 51 miles per hour, Sinclair's experiments on the same road at speeds of 50 to 75 miles per hour, Worsdell's experiments on the North Eastern Railway of England at speeds of 5 to 86 miles per hour, and Crosby's wind experiments, arrived at the formula: Lbs. per ton $= R = 2 + \frac{V}{4}$ ....(3) for speeds of over 40 miles per hour, but considered that at lower speeds the increased axle friction affected the results.

By which it will be seen that Clark's formula, fixed on years ago, is very close to the average, being slightly too high at low speeds. For the purposes of ordinary calculations the column of averages will be used. It is noticeable that at speeds of over 40 miles per hour, the wind resistance is a very large factor, while in freight hauling it is unimportant. And although passenger traffic is being carried on at higher speeds year by year, the addition of vestibules set out flush with the sides of the cars will very materially decrease that eddying around each car which forms a large part of the air resistance, while a pointed prow to the engine has been experimented on in France, and might be found advantageous at extreme speeds. The internal losses of a locomotive are, of course, not considered here, but the power as delivered to the rim of the driving wheels. The rolling friction of the engine wheels, being external to the engine, however, are included. The internal losses are about 10 per cent. (*i.e.*) 90 per cent. of the power developed may be applied to overcome the various train resistances.

(*B.*) GRADE RESISTANCES.

Grades are usually expressed as that per cent. which the rise or fall in a given distance is of that distance, or less frequently in feet rise or fall per mile of distance. Thus:—A grade of 1 in 100 = 1 per cent. grade = 52.8 feet per mile. In Figure 1 the forces which hold the

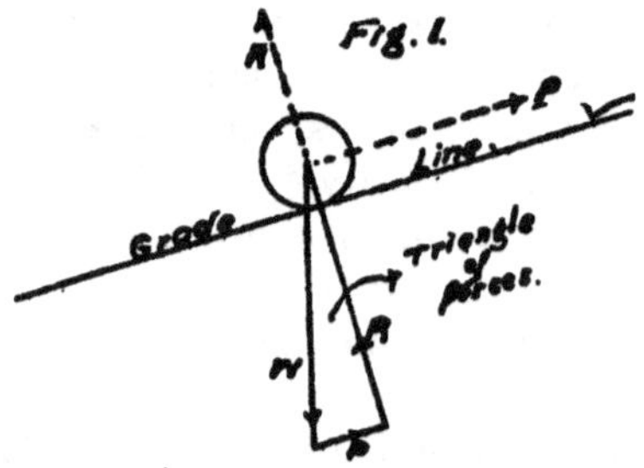

weight W, in equilibrium, are:—R, the reaction of the rails, and P the pull up the grade, the ratio of $\frac{P}{W}$ (or more precisely $\frac{P}{R}$*) is seen to be the same as the rate of grade, or, up say a 1 per cent grade, the force necessary to propel a ton at a *uniform velocity*, = $\frac{1}{100} \times 2{,}000$ lbs. = 20 lbs., and other grades in direct proportion; in other words, grade resistance is a definite calculatable quality, and varies with the grade only.

* The error in taking $\frac{P}{W}$ instead of $\frac{P}{R}$ is infinitesimal, being only about $\frac{1}{100}$ lb. out of 100 lbs. on a 5 per cent. grade.

Velocity Head on Grades.—If a train is not moving at a uniform velocity, the energy stored up in it is not constant; but varies with the speed of the train. This energy may be expressed in so many feet which the body would need to fall to acquire the velocity corresponding to a given amount of energy or velocity head $= h = \frac{v^2}{2g}$ where $v =$ velocity in feet per second, or, if we change to V miles per hour, we get $h = \frac{1}{2 \times 32} \times V^2 \times (1.5)^2 = .033V^2$ ......(4) In addition to this there is the energy stored up in the revolving wheels, which will, for ordinary cases, increase the total energy of the train in motion about 6 per cent., or change the formula to $h = .035V^2$ ....(5).

The values of h for various velocities are as follows:

TABLE VI.

V (miles per hr.)	5	10	15	20	25	30	35	40	45	50	60
h (feet)........	0.8	3.5	7.8	14.0	21.9	31.5	42.9	56.0	70.9	87.5	126.0

In considering velocity head, as it affects grades, we may treat it (*a*) as it affects stopping and starting, (*b*) as it permits of fluctuating speeds and a conservation of energy.

(*a*) The amount of force necessary to expend in starting a train, over and above what is needed to keep it moving at a uniform velocity, under the same conditions, will depend on the distance through which it acts. If, for instance, it is desired to start a train from rest and acquire a velocity of 30 miles per hour in 1,500 feet, the virtual additional grade will be $\frac{31.5}{15} = 2.1$ per cent. grade, and the additional grade resistance will be $2.1 \times 20 = 42$ lbs. per ton *to be added* to any train resistances that would exist if the train were moving at a uniform velocity. In the same way in stopping a train the brakes absorb the stored up energy more or less rapidly depending on the length in which the stop is made (*e.g.*) a stop in 250 feet from a speed of 20 miles per hour $= \frac{14}{2.5} = 5.6$ per cent. grade to which the brake friction is equivalent$=112$ lbs. per ton.

(*b*) Although it is usually considered that the grades of a road are certain definite inclines, this is only nominally so; actually, the "virtual" rate of grade against which the engine has to contend depends on what fluctuation of speed can be allowed between the foot and the top of the grade, and the length of the grade. The method of calculating "virtual" grades is as follows: Take an assumed speed, of say, 30 miles per hour at the foot of the grade, and a safe minimum speed to prevent stalling, of say 10 miles per hour at the head of the grade, then the difference of their velocity heads = 31.5 – 3.5 = 28 feet. Now this 28 feet, or whatever it may be (it may usually be limited to 25 or 30 feet), is to be deducted from the total actual grade to give the total virtual grade. It will be seen that while velocity head is a large item on short grades, it is insignificant on long ones. Thus great economy of construction can be effected often by the introduction of short, steep grades (*e.g.*) a 1 per cent. grade two miles long—actual lift = 52.8 × 2 = 105.6 feet, virtual lift = (105.6 – 28) = 77 6 feet (30 miles per hour speed reduced to 10 miles per hour at summit), and the virtual grade = $\frac{77.6}{105.6}$ = .73 per cent.

In this way on any railway profile above the actual grades should be plotted the virtual grades for different fluctuations of speed. A study of these will often enable changes in grade to be introduced, from economical motives, using short pieces of grade above the nominal maximum or dropping temporarily, in a short sag or depression, to save embankments. But the introduction of short steep grades should not be carried beyond about a 20 foot "sag" below normal; for supposing (see Fig. 2) that a train approaches *A* at a speed of 15 miles per hour, = 7.8 feet velocity head, and with uniform engine power still on, passes to *B* and then *C*, in the normal grade line, as the train passes *A* the tension on each draw bar will momentarily increase and the train will begin to accelerate its speed until *B* is reached, when the velocity will correspond to a velocity head of 7.8 × 20 = 27.8 feet, or will be about 28 miles per hour; from *B* to *C* the speed will decrease and the train will arrive at *C* with a speed of 15 miles per hour; or, in other words, the limit of "sag" is the limit which we assign to safe freight speeds (see Chapter III. for vertical curves), and may be placed at about 20 feet, defining the point up to which local depressions, in a grade, are of no importance, and may be passed over with steam on, and no brakes set, or the same brakes if descending; of course, practically, no excessive dip of very high rates of grade could be used, except temporarily during construction, for sake of appearance, and because the proper use of vertical curves would eliminate them.

It is interesting to note that although the total power exerted in passing from *A* to *C* (Fig. 2) is the same as though the train had passed along the more direct grade line, the average rate of speed is increased, and the *time* of transit from *A* to *C* is *less* on the depressed grade

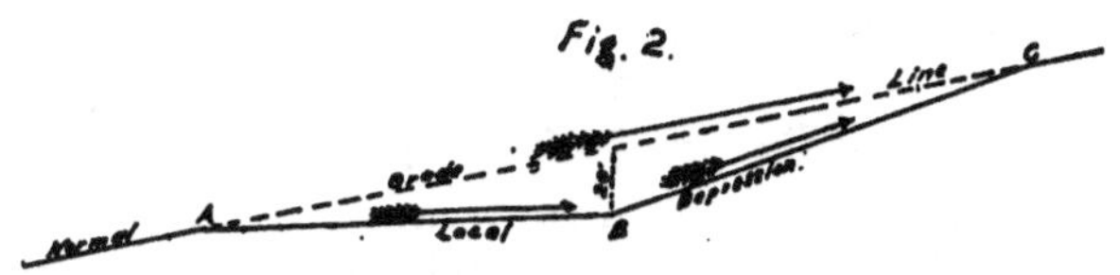

than on the direct one. But note that a rise *above* the normal grade line is not permissible, as it would probably stall a slowly moving freight train.

In connection with the fluctuation of speed, also, is the question of modification of grades at stopping places, where these grades are anywhere approaching the maximum. While level tangent resistance, at ordinary speeds of say 20 miles per hour, is about 8 lbs. per ton, it is as high as 18 to 20 lbs. per ton at zero + .. miles per hour (*i.e.*) at the instant of starting, so that any depot grade should be less than the maximum by, say, $(20-8) = 12$ lbs. per ton $= \frac{6}{10}$ grade. Also for the sake of getting up speed quickly, it should, if possible, be still further reduced, so that depot grounds are usually selected as nearly on level ground as possible. These restrictions do not apply rigidly to light passenger or local freight trains, not loaded to the full capacity of the engine, but prevent heavily loaded freight trains from stopping at stations on grades approaching the maximum.

Another danger of yards on grades is that the wind may blow cars out on to the main line from the sidings. Yards are sometimes, however, laid out with about a $\frac{6}{10}$ per cent. grade, where much sorting is done, and the cars are sorted by gravity on to different tracks.

(C.) CURVE RESISTANCES.

In America a curve is designated by the number of degrees which a hundred foot chord subtends at the centre of the circle, thus a 1° curve subtends 1°, etc. The radius of a 1° curve=5730 feet, and the radius of a

$D°$ curve is (approx.) $= \frac{5730}{D}$ feet.

The centrifugal force of a train passing around a curve at v feet per second $= C = \frac{Wv^2}{gr}$ .... (6)

If we change v into V (miles per hour) and r into D (degree of curve) we will get $C = \frac{W}{32.2} \times \frac{V^2(1.467)^2}{1} \times \frac{D}{5730} = \frac{WV^2D}{85666}$ .... (7)

Now, in order to counterbalance this force, the outer rails, on curves, are elevated sufficiently above the inner ones (super-elevation) to make the resultant of gravity and centrifugal force to pass midway between the rails and at right angles to the track, and the floor of the car will then be parallel to the track (see Fig. 3). It is evident from the figure that by similar triangles

$$\frac{\text{Super-elevation}}{\text{Gauge}} = \frac{\text{Centrifugal Force}}{\text{Weight}} \quad \text{or}$$

$$\text{Elevation} = E = G \times \frac{C}{W} = \frac{59 V^2 D}{85666} \quad \left\{\text{from (7)}\right\} \text{ ... . (8)}$$

by which it will be seen that the required elevation varies directly with the degree of curve and with the square of the velocity.

For a 1° curve, $E = \frac{59}{85666} V^2 = \cdot00069\ V^2$... .... (9)

TABLE VII.

TABLE OF SUPER-ELEVATION OF OUTER RAIL, PER DEGREE, FOR DIFFERENT VELOCITIES.

V (miles per hour) ...	5	10	15	20	25	30	35	40	45	50	60	70	80	90	100
E (inches)...	.02	.07	.15	.28	.43	.62	.84	1.10	1.40	1.72	2.48	3.36	4.40	5.60	6.90

It is evident, however, that only at that particular speed for which the outer rail is elevated will the car body be normal to the track. At slower speeds, the inner springs will compress and outer ones extend somewhat, while for higher speeds the reverse will be the case. The custom, on general traffic roads, is to elevate for medium passenger speeds of say 30 miles per hour, which is slightly over one-half inch per degree, while on high speed passenger tracks of roads having only light curves, particularly, elevations of as much as one inch per degree are common.

It may be assumed that a safe maximum riding speed will exist when the car body becomes level. Wellington's assumption is that the weight of a passenger-car will compress its springs six inches, and that the distance of the centre of gravity of the car body above the springs is equal to the distance of the springs apart giving equal turning couples. The total centrifugal force necessary for this action will be approximately

$$C = \frac{WV_1^2D}{85666} = \left(\frac{E}{6} \times \frac{W}{2}\right) + \left(\frac{WV^2D}{85666}\right) = \frac{59V^2DW}{12 \times 85666} + \frac{WV^2D}{85666}$$

or, eliminating, $V_1^2 = \left(\frac{59}{12} + 1\right) V^2 = \frac{71}{12} V^2$

$\therefore \; V_1 = 2.43V$ ....(10).

(where V_1 = speed to bring car body level.)

(V = speed for which track is elevated.)

This speed is, evidently, independent of the curvature.

The speed at which trains, running on tracks properly elevated, will overturn, is very high, and not of sufficient interest to calculate, and will depend on the amount of compression possible in the springs before the car body comes down on the buffers, and upon the amount of elevation per degree of curve given to the track.

Those roads which have sharp curves will always run at moderate speeds around them; the sharper the curve the less the speed. This fact and practical ballasting difficulties have limited the total super-elevation to about six or eight inches, which corresponds to a curve of 8° to 12°, depending on the speeds expected, on curves of greater sharpness the lessened speeds will require less elevation per degree.

The position which a short rigid truck assumes in passing around a curve is as in Fig. 4; the front outer wheel flange against the rail head, and the rear wheels radial to the curve and midway between the rails, unless the curve is so sharp or the truck so long as to render this impossible, when the rear inner wheel flange will also jam against the rail. This, however, will not happen with a

5-foot American truck on any ordinary curve. Wheels are still made with a modified coning (see chapter on track), but this is intended only to provide for future wear, as it is proven that it does not aid in passing around curves by the whole truck moving to the outside, as was formerly supposed to happen.

In passing around a curve the wheels slip in two directions, besides flange friction: (1) Longitudinal, due

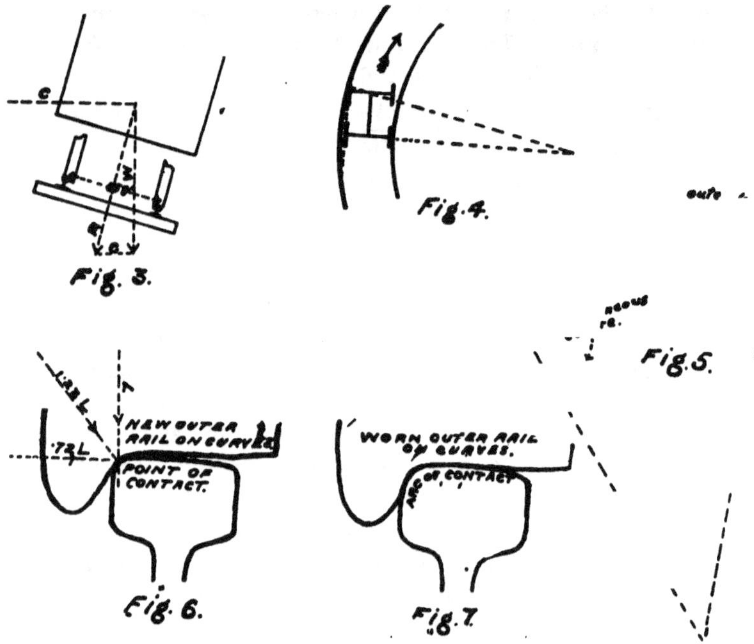

Fig. 3. Fig. 4. Fig. 5. Fig. 6. Fig. 7.

to the inner and outer rails being of different lengths; (2), lateral, due to a continuous sidewise movement in changing direction. This latter is confined to the front axle, as the rear one keeps radial always to the curve. The amount of these slippages is calculated, if necessary, as follows: (See Fig. 5.)

(1) Longitudinal slippage $= \frac{\text{gauge}}{\text{radius}} \times$ distance travelled.

(2) Lateral slippage = sin. angle a × distance travelled for 5-foot American truck, sin. a = .00087, and lateral slip = .00087 d.

It will be seen by Fig. 5 that the truck is turning continually around the inner rear wheel as a centre, and the aggregate slippages are as follows :

(1) Rear inner wheel does not slip at all.

(2) Rear outer wheel slips longitudinally.

(3) Front inner wheel slips laterally.

(4) Front outer wheel slips laterally and longitudinally.

TABLE VIII.

	1° Curve Ft.	5° Curve. Ft.	10° Curve. Ft.	20° Curve. Ft.
Average slippage per wheel, per foot........................	.00073	.00363	.00730	.0145
Average velocity of slippage (in feet per second), train going at 30 miles per hour..............	.043	.21	.43	.86
Ditto (miles per hour)............	.029	.145	.29	.54

It is noticeable that the slipping is at a very low rate, being for ordinary curves and speeds only a fraction of one mile per hour, and which ever wheel starts slipping, whether inner or outer, will continue to do so around any particular curve ; but this is immaterial.

Now, the coefficient of sliding friction between steel tires and steel rails under different velocities is about as follows :

0+	miles per hour..........	.242
7	" "	.088
13	" "	.072
34	....	.065
52	" "	.040

And as anything in the above table is nearly at zero-miles per hour, we can be safe in assuming a coefficient for this slippage at from .24 to .20, with a tendency to get less as the curve gets sharper (contrary to earlier notions on curve resistances.)

In addition to these slippages we have flange friction (see Fig. 6). When the rail is new the line of pressure is nearly a point ; the two forces acting on the front outer wheel are the load on the wheel and a lateral horizontal force sufficient to cause the slippages to take place, which have already been mentioned. If we take the coefficient of friction at .24, then there will be a total force of .72 ∠. This is combined with the vertical force ∠ to give a resultant force 1.23 ∠, acting as shown in Fig. 6. This causes the radius of the outer wheel at its line of bearing to be $\frac{1}{8}$ inch to $\frac{1}{4}$ inch larger than the inner one, and a consequent slippage takes place which is constant for all curves. The amount of friction caused by this increases rapidly as the rail becomes worn and the surface of contact increases. (See Fig. 7). So that any estimate of its actual amount will be useless unless we know the exact condition of the rail head and wheel flange.

Wellington estimates it to be 1 lb. per ton for new rails, with a considerable addition as rails become worn. This will not vary with the curvature, and is to be added to the amount calculated from Table VIII., which, taking the coefficient of friction at .24, amounts to .35 lbs. per ton per degree. The results of experiments confirm the theory that the total amount varies from $\frac{1}{8}$ to $\frac{3}{8}$ lbs. per ton per degree, increasing as the condition of the rail becomes worse, and the surface of contact, for flange friction, greater. The total curve resistance does not increase quite as fast as the curvature when the rails are new, but as the rails become worn the opposite is the case; and as the rails on curves become worn more quickly the sharper the curve, it is probably safe, as an average, for ordinary use, to assume that curve resistance varies with the curvature, and equals $\frac{1}{2}$ lb. per ton per degree. Referring back to grade and level tangent resistances, we see that—

6 lbs. per ton = resistance on 12° curve.
= resistance on $\frac{3}{10}$% grade.
= level tangent resistance at low speed.

Or that a 12° curve doubles the level resistance, or that a 1° curve is equal in train resistance to a .025% grade; strictly, therefore, we should lessen gradients on curves by .025 for each degree of curve, to compensate for curve resistance, but this would be scarcely enough, for whenever a train is nearly stalled, curve resistances due to low speed will be much higher, and if the train is once stalled it will be hard to start again on a curve just barely compensated. For this reason it is customary to compensate for curves at a higher rate than is theoretically necessary, the usual amounts being .04 to .05 per degree of curve, depending on how valuable a few feet are, in elevation, on maximum grades.

ARTICLE 7.—THE COST OF TRAIN RESISTANCES.

It must not be supposed that the expenses of operating a road vary with the total train resistances found according to the rules given in Art. 6. Even the fuel consumed is only affected partially by variation in train resistances; some are affected very much, as car repairs and rail wear, while others, such as maintenance, general expenses, etc., are hardly affected at all.

(*A*) *Cost of Curve Resistances.*—It is estimated that 36 per cent. of the working expenses vary with the curve resistances, and taking the cost of a train-mile at 90 cents, and a continuous 12° curve for a mile = 633° curvature, = one mile of level tangent resistances; then, the cost of operating a train each way for a year around 1° of curvature $= 365 \times 2 \times \frac{90}{633} \times 36$ per cent. $= 37\frac{8}{10}$ cents, therefore, we are justified in spending $\frac{100}{5} \times 37.3$ cents = $7.46 per daily train, during construction, in eliminating each degree of curvature with money at 5 per cent., any number of trains per day, or degrees of curve in direct proportion, from which it is evident that the worst features of sharp and heavy curves are more questions of appearance, comfort and safety, than of actual cost in operating, especially as the case is always one of more or less curvature only, and not a question of curves or no curves.

With roads expecting light traffic and heavy grading, it is evident that a very great amount of curvature will be justified; and then, too, it must be remembered, that it is the total angle of a curve and not its sharpness that counts in the total train resistances; the only objections to sharp curves are the slightly increased danger of derailment, the necessary slackening down from very high speeds, the slight lengthening in distance, and the sentiment of the public against them.

(*B*) *The Cost of Grade Resistances.*—Grades must be viewed from two standpoints: first, as so many feet of rise and fall, up and down which the trains must be carried; and second, as the limiting features to the maximum load which a given engine can haul at a low speed over a freight engine division. Under the first heading Wellington divides them into three classes:

(1) Those which are so light or short as to be passed over with uniform steam on and no brakes, the speed, only, fluctuating. Such grades cost appreciably no more to operate than level grades, as the trains going each way in a day gain as much energy as they lose. These grades would be, roughly, anything less than 0.5 per cent.

(2) Those in which steam is cut off in descending, but which do not need brakes in descending nor sand in ascending. It is estimated that one foot of rise and fall, per daily train per year on this class of grade, costs:

Eighty-four cents, if a minor grade, which equals $16.80 capitalized; $1.67, if a maximum grade, which equals $33.40 capitalized. These may be taken roughly as grades between 0.5 per cent. and 0.8 per cent.

(3) Those on which brakes are needed in descending and sand used in ascending. These are estimated to cost per daily train per year, per foot of rise and fall, $3.50, which equals $70, capitalized at 5 per cent. These may be taken as any grades over 0.8 per cent., unless of very short length. By multiplying the above sums ($16.80, $33.40, or $70) by the number of daily trains expected, we can arrive at the total expenditure justifiable to save each foot of rise and fall.

TABLE IX.

(See Wellington, page 544, for larger Table).

NET TRAIN LOADS OF VARIOUS ENGINES ON VARIOUS GRADES TAKING 25 PER CENT. AS THE RATIO OF ADHESION.

Grade.	Resistance per ton (in lbs.)	Passenger.		Mogul and 10-wheel			Consolidation.			Mastodon.
		Total Wt.		Total Wt.			Total Wt.			Total Wt.
		52 tons.	58 tons.	60 tons.	64 tons.	67 tons.	70 tons.	75 tons.	80 tons.	87 tons.
		Weight on drivers.		Weight on drivers.			Weight on drivers.			Weight on drivers.
		20 tons.	24 tons.	28 tons.	32 tons.	36 tons.	40 tons.	44 tons.	48 tons.	52 tons.
Level...........	8	1,198	1,442	1,690	1,938	2,183	2,430	2,675	2,920	3,163 tons
$\frac{1}{10}$ per cent....	10	948	1,142	1,340	1,536	1,733	1,930	2.125	2,320	2,513 "
$\frac{3}{10}$ " ...	14	662	799	940	1.079	1,219	1,359	1,496	1,634	1,770 "
$\frac{5}{10}$ " ...	18	504	609	718	825	933	1,041	1,147	1,253	1,357 "
1 " ...	28	305	371	440	507	576	644	711	777	842 "
1½ " ...	38	211	258	308	357	407	456	504	552	597 "
2 " ...	48	156	192	232	269	308	347	383	420	455 "
2½ " ...	58	120	149	181	212	243	275	304	334	361 "
3 " ...	68	95	118	146	171	198	224	249	273	295 "
4 " ...	88	62	78	99	118	138	157	175	193	208 "
5 " ...	108	41	53	70	84	100	115	129	142	154 "
10 "	208	0	0	7	13	20	26	31	35	38 "

Remember the above sums are not supposed to be precise, but to be as near as it is possible to arrive at the truth. These figures refer to the cost of grades regarded merely as so many feet of rise and fall, and are entirely independent of and distinct from the effect which the maximum grade has on the train load, which is a far more important matter. In special cases, as on the N. Y. C. and H. R. Railway, where the grades are very light, the curves are the limiting features, but usually grades limit and determine the load which a given engine can haul. The hauling capacity usually depends on the weight on the drivers, and the ratio of adhesion, although for high speeds the limits of boiler capacity or cyilnder power may be reached first; for freight work, however, the former are all we need to consider. The ratio of adhesion varies from 20 per cent. on slippery rails to 25 per cent. in ordinary weather, and to 33 per cent. where sand is used, but falls at once to about 10 per cent. when the driving wheels begin to slip. For any assumed ratio of adhesion it is easy to compute the load which an engine of known weight on drivers can haul up any grade. The total load includes the engine itself, but on light maximum grades it is not usual to haul maximum loads because of the difficulty in handling long trains and making couplings strong enough to transmit a very heavy pull when combined with the severe jerks caused by the great amount of slack in link and pin couplers. The increasing use of automatic vertical plane couplers having very little slack will soon do away with this difficulty and enable longer trains to be handled with facility. Table IX. enables us to compute the increased or decreased engine mileage due to a change in maximum grades, for any given amount of traffic. For light traffic such calculations must be modified, as more trains will be run to accommodate traffic than are strictly required to carry it, and only as traffic increases so as to afford at least two or three fully-loaded freight trains per day will such calculations be rigidly true—even then many roads estimate so many cars as a train load irrespective of the load on the car, and the adoption of a tonnage system for making up train loads will, in many cases, effect great economy.

It is noticeable that, on heavy grades, the level tangent resistance forms a very small proportion of the total resistance, and also that, for any given increment of grade, the increase per cent. of engine mileage is much less as the grades become heavier.

Decreased hauling capacity, on heavy grades, may be met in two ways, either by increased weight of engines, especially the weight on the drivers, as is evident from Table IX., or by increasing the number of trains (*i.e.*) the engine mileage.

The former is, of course, the cheaper method, but as the changes in grades that an engineer is called on to discuss are usually relatively small, it is only fair to suppose that the economy of heavy engines will have been realized in both cases; but supposing it possible to increase the weight of engines for heavier grades, only very few of the expenses of operating are increased. Wellington estimates that track maintenance, renewals and engine repairs are increased 50 per cent. as fast as the weight of the engine increases; fuel 25 per cent. and other items practically unaffected. Altogether, the operating expenses will only increase 14 per cent. as fast as the increase in weight of engines.

On the other hand, the usual necessary course on heavier grades will be to run more trains of less tonnage, with the same weight of engine, for a given traffic. This is a more expensive matter.

TABLE X.

INCREASE OF OPERATING EXPENSES WITH INCREASED ENGINE MILEAGE.

Item.	Cost of Item.	Per cent. increase.	Extra cost
Fuel, oil, and waste	8.8 %	67 %	5 9 %
Engine repairs	5.6 "	75 "	4.2 "
Switching engines	5.2 "	0 "	.0 "
Train wages	15 4 "	100 "	15.4 "
Car maintenance	12.0 "	– 10 "	– 1.2 "
Track maintenance, etc	17.5 "	100 "	17.5 "
Bridges and buildings	5.5 "	0 "	0.0 "
Station and general	30.0 "	20 "	6.0 "
Interest on extra engines	....	..	1.7 "
	100 %		49.5 %

It will be seen that, say, 50 per cent. of the operating expenses increase with an engine mileage increase, as compared with 14 per cent. in the first case. This is why the weight on drivers is being continually increased, and the strength of the track to carry it, on all roads having much traffic to handle, as being the cheaper expedient.

We are now prepared to estimate the cost of increasing the ruling gradient on an engine division (100 to 125 miles).

Taking a train mile to cost 90 cents, we have 90c. × 365 × 2 = \$657, as the cost of hauling a daily train (both ways) per mile, per year. If we take this *yearly train unit*, multiply it by the number of miles in a given engine division, by the increase in the number of daily trains necessitated by the heavier grades, and then by 50 per cent. (see Table X), we will have the amount which it will probably cost per year more to operate on the heavier grades than the lighter ones. If we capitalize this sum we get the amount which, for a given traffic, it would be wise to expend to construct a road with the lighter ruling grades rather than the heavier ones. (*e.g.*) To avoid changing our ruling grades from 1.0 per cent. to 1.5 per cent. on a hundred mile division, we would be justified in expending anything less than

$$\left(\frac{1000}{504}-\frac{1000}{711}\right)\times \$657\times 100\times 50\%\times\frac{100}{5}=\$328{,}500$$

for every 1,000 tons of gross freight per day, taking a seventy-five ton consolidation engine as the basis of comparison, and, roughly, two trains per day in one case and one train one day and two trains next day, in the other case, or one-half train per day difference. Now this is a very modest traffic, and yet we could afford to expend \$3,285 per mile more in one case than the other, and it is really very much more than it appears, for two reasons:

(1) Because ruling grades in most cases will probably not extend over more than one half of the road as a maximum, and we can therefore spend twice as much per mile on them, or \$6,570 per mile as a minimum, on the portions to be improved.

(2) Because all this money can be used below the ballast since track, equipment, stations, etc., in fact, all other items, remain unchanged, now to show how moderate a proportion the cost of substructure is of the cost of the whole road, the following table is given:

TABLE XI.

COST OF FOUR TRACKS OF N Y.C. AND H.R. RAILWAY PER MILE.

Grading and masonry	$22,000	=	18.9	per cent.
Bridges	3,030	"	2.6	"
Superstructure	32,500	"	27.9	"
Stations, etc.	15,400	"	13 3	"
Land and damages	15,740	"	13.6	"
Engines and cars	24,077	"	20.7	"
Engineering and incidentals	3,453	"	3.0	"
	$116,200	"	100	"

This is an extreme instance, as grading was light and equipment expensive; the items affected (1 and 2) are only 21½ per cent. of the total, and probably 25 to 40 per cent. will give a good average for ordinary single track roads. Each country traversed is suited to certain maximum gradients, and an endeavor to modify them extensively will bring very heavy additional expenses, but within narrow limits, such as a change of ruling grades by as much as $\frac{2}{10}$ or $\frac{3}{10}$ per cent., the advantages of a liberal expenditure of money to obtain the lesser grade are often overlooked and the ' penny wise " maxim adopted. Every engineer who has the decision of the ruling grade should study such figures carefully, and by as extensive surveys as possible determine what is the least ruling grade that he can get at a cost which will be justified by present or expected traffic, always, of course, considering how much money can be got at all, for no expenditure can be justified that will in any way endanger the successful completion of the road; he must consider each item of expenditure or economy, *per se*, whether it is wise or not, remembering always that it is the *difference* of gross receipts, working expenses and fixed charges that is to be thought of in determining the best general route.

Note, however, that these calculations and estimates do not hold strictly true for roads of very light traffic, because some trains must be run in any case to accommodate traffic at certain intervals, and if they are not fully loaded, then an increase of grade will not have any effect until it causes an increase in the number of trains, as a change in the rate of grade does not usually mean any increase in the total rise or fall.

In comparing two routes for costs of operation the best method is to assemble the curves and grades of different classes and take their differences, pro or con, also the difference in the number of trains per day necessary to handle the probable traffic. These differences multiplied by their proper multipliers will give a comparison of how much more valuable one route will be than the other for a given traffic, and will determine consequently how much more can be justifiably spent to construct one route rather than the other, *other things being equal.* In such a comparison it will be found that any difference in the ruling grade is usually the preponderating item.

CHAPTER III.

CURVES.

ARTICLE 8—VERTICAL CURVES.

Wherever there is a change in the rate of grade there must be a vertical angle or a vertical curve. If this change is slight, less, say, than $\frac{3}{10}$ feet per 100 feet, no need exists, either on construction or afterwards, of doing anything more than to let the trackmen put in a slight curve by eye, but when the change is of considerable magnitude, care should be taken, both for the sake of appearance and also for safety, that a regular vertical curve unites the two grade lines.

In the past, in America, this has not been often done. If ascending and descending grades were to be united, a short piece of level grade was inserted at the summits and in the depressions; anything further was, curiously enough, relegated to the track gang as being a refinement unnecessary for a civil engineer to bother with; the track or section foreman, with greater appreciation of the real need for a regular increment of change from one grade to another, did the best he could and put in vertical curves by eye, which moderated the ill-effects of such neglect. Wellington has ably dealt with the subject, at length, from the standpoint of the link-and-pin coupler, and demonstrates that the vertical curve which is needed, theoretically, is one which will change the rate of grade from the front to the rear of the longest trains run over the road by an amount not greater than the grade of repose (the grade of repose is that grade down which a train will just keep moving under its own weight, and is about $\frac{5}{10}$ per cent. for loaded trains at a speed of 25 miles per hour, and increases with the speed). He reasons thus: Taking the train as a whole, each car will momentarily crowd toward the one in front of it, and so on throughout the whole length of the train, putting it in a state of compression, with slackened couplers if the grade resistance at the front of the train is enough greater than at the back end to exceed the grade of repose. This is based on an assumption of uniform engine power, and should the engine driver increase speed just at this

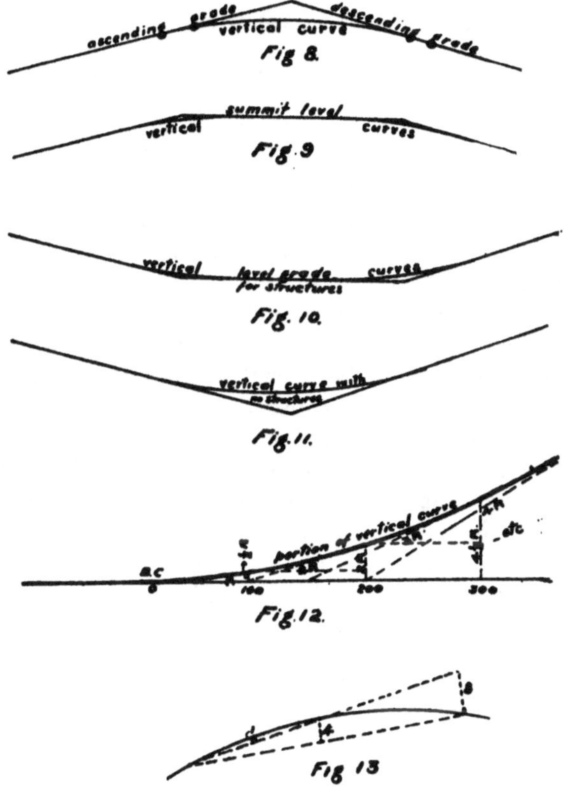

Fig 8.

Fig 9

Fig. 10.

Fig. 11.

Fig. 12.

Fig 13

instant, when everything is slack, the tendency will be to create severe jerks and oscillations causing derailments. This reasoning refers entirely to a grade depression, whereas at a summit the reverse will happen and the couplers will be momentarily strained much more than normally. From these premises we can see that the vertical curve at summits may be arbitrary in amount and much sharper than in depressions. Probably a change in rate of grade of $\frac{2}{10}$ per cent. for each 100 feet is not excessive, and may be inserted either as a complete curve joining the ascending and descending grades (see Fig. 8), or if the summit level is long it may be divided into two portions (see Fig. 9). When, however, a descending grade is to be united to a level or ascending grade, an accurate calculation should be made for reasons already given. For instance, supposing that the longest train on the road will be 500 feet (engine and 14 cars), then

$$\frac{\frac{5}{10} \text{ per cent.} \times 100}{500} = \frac{1}{10} \text{ per cent. change}$$

per 100 feet will be the amount strictly demanded for complete safety on a road of the given length of train using link and pin couplers. But as automatic vertical plane couplers, with practically no slack, come more generally into use, which is only a question of a few years, the need for such extensive curves will not be imperative, and a vertical curve changing more rapidly will answer fully, when a longer curve is difficult to obtain. Usually, however, a level grade between the descending and ascending grades is required, because a structure should always be placed on a uniform grade from end to end, and as they are usually in the depressions, this limits the vertical curve in such cases to two short pieces joining the level grade to the others. (See Fig. 10.) If there is no break in the embankment a continuous vertical curve is much better from every point of view, and should be put in as in Fig. 11.

On roads having only light grades, and consequently heavier and longer trains, the rate of change in depressions will be very small, and circumstances will determine whether the full amount can be put in without excessive cost; but with light grades and easy vertical curves, the distance which the middle of the curve will rise above the point of intersection is small. It may be calculated, in any case, in the same manner as the middle distance in horizontal circular curves, if the vertical curve is treated as a circle, or if treated more precisely as a parabola, it may be stated at once as half the distance which the apex is from the middle of a chord drawn from one end of the vertical curve to the other, this being a fundamental property of the parabola.

(*a*) Treating the vertical curve as an arc of a circle, calculate first the permissible change in grade per 100 feet divide this into the total change of grade, giving the total length of curve, ½ of which will be on each side of the apex of grades; then the position of the curve for each 100 feet relatively to the tangent lines may be obtained graphically on a large scaled drawing, or calculated more precisely as in ordinary horizontal circular curves.

(*b*) Treating the vertical curve as a parabola having a constant rate of change of direction per 100 feet, is more precise and more convenient. Calculate first the length of curve, which will be the same as in (*a*), and then proceed as follows: Let the change of grade per 100 feet = R. Then referring to figure 12—, the departure of the curve from the tangent will be ½R, 2R, 4½R, 8R, 12½R, ... etc., till the middle of the curve is reached, after which the distances from the second tangent will recede . . . . 12½R, 8R, 4½R, 2R, ½R, to the other end. It will be seen that by the latter method the elevations are always in even units or portions of units, and the rise of the curve above the tangents is given almost by inspection; for convenience the length of a vertical curve should be fixed at the nearest *even* hundred feet, so that the curve may be divided into two equal parts of exact hundreds in length. Such vertical curves, with their elevations once established, will be no more difficult to place on the ground and build to than a succession of straight lines with abrupt changes in grade, and will give a track safer in depressions, having better drainage in summit cuts, and better in every respect, but increasing the cost of the road-bed slightly.

ARTICLE 9.—HORIZONTAL CIRCULAR CURVES.

It is not necessary to treat here of the mathematics of the circle. There are several engineering field books which have considerable space devoted to methods of placing curves on the ground under ordinary or exceptional circumstances. Some of these books also contain, in addition to ordinary mathematical tables, tables of external secants and of sub-tangents for each degree of curve, and for each interval of one minute in the total intersection angle; these books are great time-savers in field operations, and should always be used.

In placing curves on the ground, it is preferable to establish the two tangents first, intersect them and measure in the BC and EC from the intersection or apex; then the curve can be run in from either or both ends and any error minimized. With very long flat curves on unstable ground, it may even be preferable to fix the middle of curve from the apex by measuring in the external secant, and then run the curve in from the ends and middle; the method sometimes adopted of running a curve in from the BC, and deflecting on to the second tangent at the EC, is very liable to establish it erroneously.

Another very important point is the method of keeping curve notes. The vernier should always read half of the total deflection of the curve from the BC up to the point on the curve toward which the telescope is pointing; this is a constant index of the position of any point. This method necessitates loosening the vernier-plate at each set-up and re-setting it to read the index reading of the back-sight; but it has the all important feature of enabling a transit to be set up at any point on a curve, and being sighted to any other point with a certain knowledge of what the vernier reading should be. Curves can be run in backward as easily as forward. Any other method of keeping notes will be found, in the end, less reliable and convenient. Whenever curves are sharper than 4° or 5° it is better to put in stakes every 50 feet even on easy ground, as the difference between the length of chord and curve for 100 feet measurements would be considerable; it is also convenient for cross-sectioning. In running in sharp curves, particularly curves having a large intersection angle, the greatest care is necessary in the chaining; poor results in checking up at the EC are usually traceable to the errors in measuring the subtangents or the curve itself.

It is often necessary to replace stakes that have been lost, or to put in intermediate stakes on curves without the aid of a transit; whenever this is the case it is valuable to remember the following formula, which is approximately true for all curves usually used on steam railways:

It is $O = .218\ N^2D$...... ....(11)

Where O = offset from middle of a chord to the curve (in feet)

N = length of chord in 100 feet.

D = degree of curve.

Or, if simpler, remember that the offset from the middle of a 100-feet chord on a 1° curve is .22 feet, and that

(1) Offsets vary directly as the degree of curve.

(2) Offsets vary as the square of the length of chord, which is true up to 200 or 300 feet chords.

(3) Offsets, inward to a curve, from a prolonged chord are 8 times the offsets from the middle of the same length of chord outward to the same curve. This is illustrated in Fig. 13.

Circular curves are in general use on railways, but there have been isolated attempts at using the parabola, which have not been found satisfactory. The idea involved in its use was to have a curve of easy radius at the ends and sharper in the middle, but the train did not travel steadily, being in a constant state of change from beginning to end of curve. It has been found from the very first days of railroads that an annoying and dangerous jolt, sidewise, took place as a train either entered or left a curve, and the parabola was a first or rather a mistaken idea as to remedying this evil.

Instead of this, the concensus of opinion has fixed itself on the use of the circular curve, but with the modification of the use of easement curves at each end of it to join it on to the tangents in such a manner as to modify or wholly dissipate any disagreeable shock which would occur if the curve were to change instantaneously to a straight line. In the past the trackmen have been allowed to introduce these easements themselves in an approximate and makeshift manner, but at present there is a growing feeling that an accurately calculated and placed easement curve is necessary, especially as passenger speeds are becoming higher. Easement curves have been used for many years in Europe, and are becoming quite common in America.

ARTICLE 10.—EASEMENT ON TRANSITION CURVES.

In article 6, under "Curve Resistances," is given formula (8), which indicates the amount that the outer rail on a curve should be elevated above the inner one, but, as the two rails on the adjoining tangents are of the same height at any given point, the question arises as to the best manner of effecting this change of conditions so as to lessen any shock to passengers or rolling stock, or indeed to entirely abolish it. Practice has determined that, where there is distance enough to permit, the curve super-elevation ought not to be lowered more than ½ inch per rail length (30 feet), or (*e.g.*) on a 10° curve of ½ inch super-elevation per degree, this would require a distance of 300 feet. The most common practice in America has been to bring the full elevation to the ends of the curve, and then lower it on the tangents. This, evidently, will act so that as a train approaches a curve the play of the wheels (½ inch to 1 inch) will all be at the outside, *i.e.*, the wheels will press against the inner rail, and then, at the instant the curve is reached, there will be a lurch to the outside in assuming the natural position, in passing round a curve, of the front wheel of each truck against the outer rail.

Some have tried to remedy this by lowering the elevation partially on the curve, and partially on the tangents, which merely divides one shock into two smaller ones. The true remedy lies in not making an abrupt change in horizontal alignment from a curve to a tangent or *vice versa*; but in so arranging the track at each end of a curve, that commencing with a curve of infinite radius, this radius is gradually decreased, *i.e.*, the curve is sharpened, and at the same time, the elevation of the outer rail is increased, keeping this elevation at each point just sufficient for the curvature until a junction is made with the main circular curve, with a curvature equal to it, and with a full elevation, and having kept an equipoise between curve and elevation at each instant, all lurches and shocks will be avoided. That this is the only true and rational solution, is proven by the fact that practical trackmen, unguided and even hindered, often, by engineers' rigid centre stakes, but recognizing the evil and its remedy, have introduced crude easement curves wherever they could do so, and improved the situation as much as possible; but as the tangent and main circular curve were both fixed in position by construction, all that could be done was to flatten the ends of the curve at the expense of the adjoining portions, which were thus made sharper than the main curve itself, and formed more or less of elbows in the track, often 2° or 3° sharper than the main curve.

Now this can be avoided by moving the curve inward bodily, or by changing the position or direction of the tangents, or by sharpening the whole curve slightly, any of which will permit of the introduction of proper easement curves at the two ends of the circular curve. Many methods have been advocated for putting in these easements, the endeavor being to simplify the process, in point of time and mental effort, and still preserve the essentials. Some of these are: (*a*) A succession of short pieces of curves of decreasing radii. (*b*) A modification of (*a*) in the form of a spiral. (*c*) A modified quadratic parabola (Holbrook spiral). (*d*) A modified cubic parabola. As any one of these can, when once understood, be easily laid out in the field, it is only necessary to decide on the most adaptable and suitable one for all cases to be met with, and study its theory and actually use it, after which its seeming difficult nature and laborious methods of application, so long dreaded by many railway engineers, will be found quite simple, and capable of rapid manipulation.

Almost all engineers are agreed that transitions are intrinsically necessary, and on European and the best American tracks their use has become established; the chief objections to their general adoption here have been the deeply rooted ideas that they were difficult to apply and too refined for ordinary use, but as speeds are being increased and competition is keener, they are beginning to be used by all roads of any importance because the consequent easier riding caters to the travelling public and also because the wear and tear on the rolling stock, and the difficulty of keeping the ends of curves in proper line, are thereby much decreased.

(*a*) This first class of transitions does not require any demonstration. Some engineers put 100 feet or 200 feet of a curve of larger radius at each end of the main curve, and trust to the trackman for the rest, others introduce a series of short arcs of decreasing radii, say 30 feet of 1° curve, 30 feet of 2° curve, etc., leading up to the main curve at the rate of 30 feet per degree; this necessitates placing the transit every 30 feet, is a tedious and clumsy method, and the result is that the trackmen fuse one portion into another until it is, to all intents and purposes, the same as a spiral. It does not admit of ordinary calculation or manipulation unless modified as in the next paragraph.

(*b*) In "The Railway Spiral," by Searles, is given a complete analysis of the transit work necessary to lay down a succesion of short circular arcs, beginning at zero, and having equal lengths of arcs of equal increments of sharpness, *e.g.*, 20 feet of 1° curve, 20 feet of 2° curve, etc., up to any required sharpness. Tables of deflections are worked out, so that any point of change of curvature can be used as a transit site, and any point of change can be established from any other point of change by transit deflections. Methods of conversion are also given, so that from one foundation series other deflection tables may be determined suitable for spirals of more or less rapid sharpening. The subject is well discussed and thoroughly worked out for all probable conditions, but as it does not present that same flexibility and simplicity of use which the cubic parabola possesses, its continued use is doubtful. It has served its day, and, where used, furnished the trackmen with a succession of hubs, really the ends of arcs of increasing sharpness, but practically points on a spiral very suitable for an easement curve.

(*c*) The Holbrook spiral (quadratic parabola). The idea involved in this easement curve is that the *vertical* acceleration of the train, as it passes around it, should be uniform. If we let t represent horizontal distances (with train moving at a uniform speed) in the general formula $s=\frac{1}{2}\text{ft.}^2$, then, in order to keep f (acceleration) constant, the distance, s, (*i.e.*) the amount which the train rises above the normal tangent level, must vary as the (distance)^2, and as the elevation should always bear a constant ratio to the degree of curve at each point, therefore the degree of curve on this required spiral must vary as the square of the distance from the zero of such a curve, (*i.e.*) the radius of curvature, at each instant, must vary *inversely* as the (distance)^2 from the zero of the curve.

A curve of such a nature has the equation $y=(f)x^4$ to represent it, and is a curve very flat at the beginning, but increasing very rapidly in curvature. This easement curve sacrifices the correct horizontal alignment, as will be seen in the next paragraph, for a supposed refinement in the vertical one; it is quite difficult to apply except in most ordinary cases, as the formulæ used involve expansions of sine and cosine, does not present any advantage over the cubic parabola, and is not so adaptable or easy to manipulate in the case of any problems having special conditions.

(*d*) THE CUBIC PARABOLA.

This curve as adapted to transitions to railway circular curves has been studied pretty thoroughly. Howard, Armstrong, and others have written pamphlets on it; the transactions C.S.C.E. for 1891, 1892 and 1893 have several papers and discussions on it, and its probable originator, the late A. M. Wellington, determined very simple equations for it which were published in the *Engineering News*, January and February, 1890.

It is this last demonstration that will be now given to which will be added necessary developments. The curve required for a suitable transition is one which starting with an infinite radius or D (degree of curve) $= O$. at the $B\ T\ C$ (A Figs. $13\frac{1}{2}$ and 14) has a degree of curve at each point in direct proportion to its distance from the $B\ T\ C$ until it joins and becomes tangent to the main curve at C, and is, at that point, of the same degree of curvature as the main curve.

The cubic parabola $y = (f)\ x^3$ approximates to these conditions.

Let $A\ M\ C$ (Fig. $13\frac{1}{2}$) be the cubic parabola, $A\ C^1$ tangent to it at A, and $I\ C$ the radius of the D degree curve with which it connects at C, having there a common tangent $I^1\ C$.

Let X be the central angle of the circular arc $P\ C$, which is changed into the transition curve $A\ M\ C$.

Let $E\ P\ G$ be tangent to $P\ C$ at P and therefore parallel to $A\ C^1$, and make $C\ C^1$ perpendicular to $A\ C^1$.

Also in Fig. 14, let vertical heights represent degrees of curvature at any point and horizontal distances, measurements along the cubic parabola.

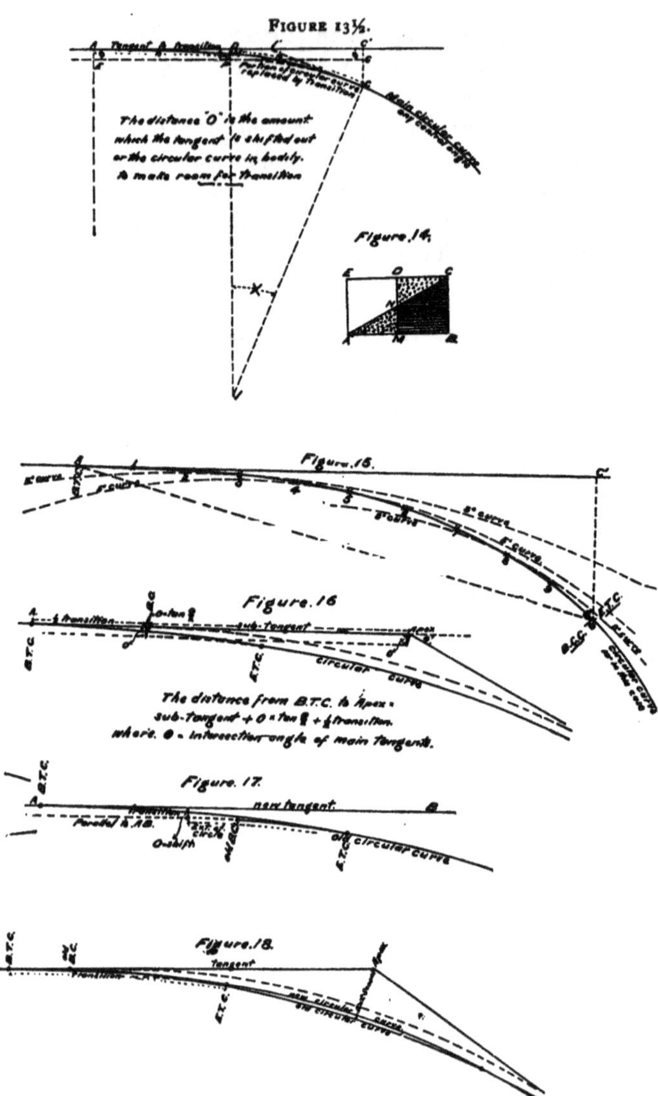

Then the rectangle D B will represent graphically the circular arc *P C*, and the triangle *A B C* represent graphically the cubic parabola *A M C*, and from this diagram and Fig. 13½ we may readily conclude :

(1) Because the total angles of the arc *P C* and the transition *A M C* are equal, therefore the area of the triangle *A B C* must equal the area of the rectangle *D B*, and

therefore a transition curve is always twice as long as that portion of the circular curve which it replaces.

(2) Because the triangles $A\ M\ N$ and $C\ D\ N$ are equal and similar, therefore the angular deflections or offsets from the tangent to every point in $A\ M$ (Fig. 13½), and from the circular curve outward to every corresponding point in the equal distance $C\ M$, are equal in magnitude and distribution, and $\therefore$ $D\ M$ is equal to $M\ P$ and half of $D\ P$ (Fig. 13½). Hence the offset or shift $D\ P\ (= O)$, and the transition curve $A\ M\ C$ bisect each other at M.

(3) The offsets from a tangent to a circular curve vary as the square of the distance from the tangent point (nearly), or to formulate it O varies as $n^2\ D$.

Where O = offset from tangent,

n = distance of the offset from the tangent point,

D = degree of curve,

but by our definition of a transition curve the degree of curve at any instant also varies as n. Therefore in a transition curve of this nature O varies as $n^2 \times n = n^3$......(12), and also by paragraph (2). If we have a given offset from tangent to circular curve at $D = D\ P\ (= O)$, then the offset to the transition curve at a distance m from A is equal to $\left(\frac{m}{n}\right)^3 \frac{O}{2}$ where n = ½ length of transition = $A\ M$ = $M\ C$. And, in the same way, measuring back from C along the curve toward P at any distance, m the offset outward from the circular curve to the transition curve =

$$\left(\frac{m}{n}\right)^3 \frac{O}{2} \qquad \text{(13)}$$

The equation to the cubic parabola can now be established in terms of the offset O and ½ length n.

let $y = C\ x^3$, but when $y = \frac{O}{2}, x = n$, therefore $C = \frac{O}{2} \times \frac{1}{n_3}$

$$\text{and } \therefore y = \frac{O}{2n^3}\ x^3 \qquad \text{(14)}$$

(4) because by equation (12) offsets to a transition curve vary as cube of distance from origin, therefore in Fig. (13) $C\ C^1 = 8 \times D\ M = 4\ O$, and therefore $G\ C = 3\ O$ ....(15)

Now, for very small angles, $G\ C = P\ C \times \text{Sin}\ \frac{X}{2}$ (nearly), and $P\ C = 2 \times I\ C \times \text{Sin}\ \frac{X}{2}$ (nearly), therefore by substitution we get

$G\,C = 3\,O = 2\,I\,C \times \text{Sin}^2 \frac{X}{2}$ (nearly), but $I\,C = \frac{5730}{D}$ (D = degree of curve), and $\therefore\ O = \frac{3820}{D} \times \text{Sin}^2 \frac{X}{2}$, and

$$\text{Sin}\,\frac{X}{2} = \sqrt{\frac{O \times D}{3820}} = .01618\,\sqrt{O \times D} \ldots\ldots\ldots\ldots(16)$$

from which we can get X, having O and D; or otherwise, since $n = \frac{X}{D}$ (evidently), and for small angles $\text{Sin}\frac{X}{2} = .0087\,X$ (in degrees).

$\therefore$ substituting in (16), we get

$$X = \frac{.01618}{.0087}\sqrt{O \times D} = 1.86\sqrt{O \times D} \ldots\ldots\ldots\ldots(17)$$

and $n = \frac{1}{2}$ length of transition $= 1.86\,\frac{\sqrt{O \times D}}{D} =$

$$1.86\sqrt{\frac{O}{D}} \ldots\ldots\ldots\ldots(18)$$

This can also be put in the approximate form,

$$O = \frac{n^2}{6R} \ldots\ldots\ldots\ldots(19)$$

Where R = radius of the curve.

Equations (12) to (19) give such relations between X, n, O and D as will enable any length of transition curve to be put in for any degree of curve.

(*e.g.*) Let $D = 10°$ curve, and $O = 10$ feet.

Substituting in (18) we get, $n = 186$ feet, or the transition is $2n = 372$ feet, which is somewhat longer than is needed.

(*e.g.*) Let $X = 15°$, and $D = 10°$ curve.

Then, $O = \frac{15 \times 15}{(1.86)^2 \times 10} = 6.5$ feet, and $n = 100 \times 1.86\sqrt{\frac{6.5}{10}} =$ 150 feet, which latter could have been determined directly.

Also $\frac{O}{2} = \frac{6.5}{2} = 3.25$ feet, and any other offset will vary as cube of distance from A; that at the quarter points being, for instance, $(\frac{1}{2})^3 \times 3.25 = .41$ feet.

A most usual length of transition is 30 feet per degree of curve, which permits of the super elevation being lowered at $\frac{1}{2}$ inch per 30 feet = 1 rail length, which is a most usual amount.

Now, although these equations enable us to put in transitions by offsets, if we have for instance, the tangents already in place, and can move the main curves inward bodily so as to permit the requisite "shift" O, which is very useful if, on construction, the rigid curves and tangents are found already in place, and offsetting is the quickest method to use—still, we also wish to be able to put in transitions as a regular part of location, and not as an afterthought, and to do so it is necessary to determine methods of locating such curves by transit deflections from the beginning, end, or intermediate points.

Any small angular deflection from a meridian to any point varies as $\frac{\text{offset}}{\text{distance}}$, or in other words the natural tangent of any small angle is its circular measure.

Now referring to equation (12) and Fig. (13½) any offset from the tangent $A\ C$ to the transition curve varies as the cube of the distance from A.

∴ angular deflections to the transition curve from tangent $A\ C$, using A as origin, vary as $\frac{\text{offset}}{\text{distance}} =$ $\frac{(\text{distance})^3}{\text{distance}} = (\text{distance})^2$ .... (20)

Also in Fig. (13) $\frac{G\ C}{A\ C} = \frac{1}{2}\ \frac{G\ C}{P\ C} = \frac{1}{2} \times \frac{X}{2} = \frac{3 \times O}{2 \times n}$ (evidently)

∴ $X = \frac{6 \times O}{n}$ but the angle $C'AC = \frac{C'\ C}{C\ A} = \frac{4 \times O}{2 \times n} =$ $\frac{2 \times O}{n}$

∴ the angle $C'\ A\ C = \frac{1}{3} \times \frac{6 \times O}{n} = \frac{1}{3}\ X$ ...... (21)

Equations (20) and (21) enable us to determine any deflections to the transition curve from the point A; (*e.g.*) let a 10° curve have a transition curve 300 feet long then $X = \frac{n}{D} = \frac{300}{2} \times \frac{1}{10} = 15^\circ$.

∴ by (21) the angle $C'\ A\ C = 5^\circ = 300' =$ deflection from tangent at A to the end of the transition, and by equation (20) the deflections to each $30'$ intermediate point are:

1st 30 ft. point $\left(\frac{30}{300}\right)^2 \times 300' = 03'$

2nd $\left(\frac{60}{300}\right)^2 \times 300' = 12'$

3rd $\left(\frac{90}{300}\right)^2 \times 300' = 27'$

4th $\left(\frac{120}{300}\right)^2 \times 300' = 48'$

etc., etc., etc., etc.

This series of deflections from the origin A, continued as far as necessary, may be called *a foundation series*, and is the basis of all deflections forward or backward from any point. We must now, in order to fix on intermediate deflections, with the transit also at some intermediate point, look on a transition curve, thus: (See Fig. 15). Suppose it to be stopped at 1, then it is a transition curve to a 1° curve; if stopped at 2, it is a transition curve to a 2° curve, etc.; therefore if the transition *does* continue past the points 1, 2, 3, etc., we may consider it to be composed of two parts: 1st, a 1°, or 2°, or 3°, etc., curve, according to circumstances. 2nd. Plus the *foundation series* of 3′, 12′, 27′, 48′, etc., beginning at the point considered, and continuing forward to any desired extent, and the transition curve deflections are the sum of these two. Also, in the same way, the transition curve deflections looking backward, with transit at any point, are those of a certain degree of curve corresponding to this point, minus the same *foundation series;* (e.g.) suppose the transit to be at the point 3, with the vernier at zero, and line of sight tangent to the curve, then the vernier readings to each intermediate point would be—

$(A) -\left(\frac{90}{100}\times\frac{180'}{2} - 27'\right) = -0^\circ\ 54'.$

$(1) -\left(\frac{60}{100}\times\frac{180'}{2} - 12'\right) = -0^\circ\ 42'.$

$(2) -\left(\frac{30}{100}\times\frac{180'}{2} - 3'\right) = -0^\circ\ 24'.$

(3) $0^\circ\ 0' = 0^\circ\ 0$ position of transit.

$(4) +\left(\frac{30'}{100}\times\frac{180'}{2} + 3'\right) = +0^\circ\ 30'.$

$$(5)+\left(\frac{60}{100}\times\frac{180'}{2}+12'\right)= +1^\circ\ 06'.$$

$$(6)+\left(\frac{90}{100}\times\frac{180'}{2}+27'\right)= +1^\circ\ 48'.$$

$$(7)+\left(\frac{120}{100}\times\frac{180'}{2}+48'\right) = +2^\circ\ 36', \text{ etc.}$$

In this way a table can be prepared giving deflections to be made to any point (every 30 feet), with transit located at any point. These tables are conveniently made out by Mr. Armstrong, for 30-foot chords = 1 rail length; but different foundation series and different tables may be made out, or special calculations made by equations (12) to (20) for a transition curve of any rapidity of sharpening, but of the same nature and handled in the same way. This is often necessary where there is not room between the *BC* of one curve and *EC* of the previous one to permit of the introduction of transitions which sharpen so slowly as 30 feet per degree.

In street railway work, for instance, transitions sharpening from 0° to 20°, or even 40°, etc., are needed, and must not occupy more than 20 or 30 feet in length. Special corrections must be applied in such a case, and even for steam railways Mr. Armstrong has worked out corrections in lengths to apply to the very approximate equations here given, but as the correction is zero until an 8° curve is reached, and only 1 foot in 300 for a 10° curve, it is hardly worth taking account of here. Any one desiring extreme accuracy for curves from 8° upward, are referred to J. S. Armstrong's pamphlet.

The three problems most frequently met with in practice are briefly as follows:

1. (See Fig. 16.) To keep tangents fixed and to move the circular curve inward, retaining the same degree of curvature. In this case, take an arbitrary offset or length of transition, and determine the other unknowns by foregoing equations. The distance from the apex of tangents to the *B T C* consists of three parts:

(*a*) Sub-tangent of circular curve $= R \times \tan\frac{\theta}{2}$ (R = radius).

(*b*) Correction of shift = $O \times \tan\frac{}{2}$. (See Fig. 16).

(*c*) ½ length of transition = *n*.

The amount in (*b*) is usually very small, unless is large.

2. (See Fig. 17.) To keep the circular curve fixed, and move out the tangents either in direction or position, or both: If the tangents are moved outward and kept parallel to their original positions, proceed as in (1), except that the correction of shift (*b*) does not exist. If the tangents are not moved outward parallel to their original positions, but pivoted about some distant point, then calculate the angle pivoted, and continue the circular curve through an equal central angle. So that a tangent to the curve at the new *B C* or *E C* would be parallel to the pivoted tangent; then measure the amount of shift *O*, and by the ordinary equations calculate the unknowns; the amount of shift *O* could be calculated without any field work. No correction of shift is here necessary; this second case is most usually met with in revising location, and is very convenient often in the final slight movement of tangents or curves, by avoiding the running over again of the whole circular curves, often situated on a rough hillside or heavy bush, and yet enabling a tangent to be moved on to better ground.

3. (See Fig. 18). To sharpen a curve and introduce transitions, so that the track will not be altered in length; this problem is the one met with in re-running old track centres where transition curves have not been previously used.

The method of solution is to assume an external secant slightly less than the original one, by an amount = expected shift, *O*, + an arbitrary amount of five inches to ten inches, depending on the sharpness and total central angle of the circular curve; then calculate the transitions and complete position of a curve of assumed external secant and given total central angle, and, either by plotting or calculations, determine whether this new curve will cross the original one about at the ¼ points and give the same length of track, thereby minimizing the movement of the track. If in error, a second trial will give usually satisfactory results. This method will often be found to give transitions, which, unless the central angle is large, will occupy the whole central angle, leaving no circular curve at the centre. As this is not desirable, it is preferable in a case of this kind to use shorter and sharper transitions, so as to retain a considerable portion of circular curve at the centre.

While these are the three usual problems to solve, others may arise such as introducing a transition at a point of compound curvature which needs special solutions. For further details, the reader is referred to the literature already mentioned, and the engineer, young or old, who has not used transitions in the field, is advised to become familiar with some one of the forms given, and actually put it into practice, when its seeming tediousness and difficult nature will disappear.

He should recognize that, as he would be quite ready to spend a few hours extra now and then, during railway construction or maintenance, on trivial matters such as affect the general appearance of the road only, and are not really important, he should be far more willing to give much additional labor and attention to such a question as this, when the returns will be increased comfort to travellers, decreased wear on rolling stock, and greater ease in retaining good alignment at the ends of curves. Whenever transitions have been used, their beneficial effects have at once been recognized, and, once established, trackmen maintain them easily and instinctively. Some of the oldest and most conservative of the American roads are now engaged in introducing them on their main tracks.

CHAPTER IV.

ARTICLE 11.—SURVEYS.

The final determination of the exact centre line of a railway roadbed and track is only reached after a process of sifting, which extends from the first thought of the necessity for such a railway until the track is laid. Roughly speaking, it is usual to divide the operations into three stages, which, however, often overlap each other, or are again divided into subsidiary steps. These customary general divisions are:

(1) Reconnaissance.
(2) Preliminary or Trial Line Surveys.
(3) Primary and Revising Location Surveys.

ARTICLE 12.—RECONNAISSANCE.

Reconnaissance may be said to begin after it has been decided that there is a necessity for a railway between two given terminals, or along a given route.

In the latter case, local considerations, or the shortness of the distance, or the existence of a definite water line route, may limit the scope of explorations, but looking to the larger problem, where an engineer has to determine what is the best route between two terminals several hundred miles apart, the study is interesting and one requiring a high order of talent. If the country to be traversed is unsettled, or thinly settled, the problem is simplified by lack of railway competition often, or even by considerations of traffic, but it then demands a close investigation of the natural resources of the country, which, though dormant, will be developed by the railway itself, and it might be considered best, all things considered, to build sometimes, at a sacrifice of distance, grades, or capital outlay, through a country of great natural resources, rather than through a barren one by a route physically superior. On the other hand, through a populous country, the question is much more complex, by reason of the existence of other railway routes already established; but, on the other hand, simplified by a more or less well defined trend of population, which indicates the probable future distribution of people in accordance with natural laws. For these and many other reasons, exploration should commence and be well under way, or even completed, before instrumental work commences; it should, at least, be completed for such a distance that some critical place has been reached through which the final location must pass.

In order to finally fix on the best route between two defined points, it is necessary to study a wide belt of country; even a great number of trial routes will not answer so well, because portions of various routes may be finally selected and joined together. In order to explore such a wide belt of country, use must be made of all existing maps. These when made from governmental surveys will be found of extreme service as a skeleton on which to build such additional information as may be necessary to complete the study in hand. All streams, summits, passes, etc., within the extreme margin of possible routes should be accurately fixed in plan and elevation. A knowledge of the classes of timber, stone, and excavations, and of difficult river crossings, etc., should be included, and from such data, together with closely estimated lengths of lines, ruling grades (obtained from barometer heights), probable traffic, cost of construction, difficulties of maintenance and dangers of future or present competition, a selection is made of the two or three most favorable routes, over which it is thought necessary to make instrumental surveys.

In carrying out reconnaissance, the instruments required will depend on the class of work to be done. These should always include an aneroid barometer, a Locke level, a pocket or prismatic compass and a field glass; distances may be determined from maps, if existing, by pacing, by the rate of travel of a horse, or if in open country, it will be better to take the time to determine them by stadia or some form of telemeter. The aneroid barometer is an instrument supposably compensated for temperature, and under static air pressures capable of always reading the same at the same altitude;

but errors in graduation, in workmanship and adjustment, and the barometric changes going on in the atmosphere make it far from a precise instrument. In order to make it available, each instrument when purchased should be rated alongside a mercury barometer, and only those which have a reasonably uniform and small rate of error should be accepted, so that a table of such errors can be prepared and used in conjunction with actual readings taken. Aneroids in high altitudes are often much in error, and generally speaking, should be used to obtain *differences* in elevation rather than actual ones. If a barometer is read at the same spot every hour for a day, a continual fluctuation will be noticed, even during bright dry weather and very much more so during periods of storm or change; these readings if plotted may be termed the ***diurnal gradient.*** It is evident, therefore, that readings from an aneroid taken at various places, at different times, even during the same day, will not be reliable, and in order to make such readings of value, there should be another stationary aneroid read at regular intervals, and the readings of the moving aneroids corrected according to the fluctuations observed at the central point. Should only one aneroid be available, it would be better, where possible, to make two or more determinations of the same points at different times, to get an average, and to work only when the atmosphere is in a settled condition. Equipped with the above-mentioned instruments and one or two assistants, the engineer on reconnaissance should go into the field free from prejudice; the well-known wagon road or trail may be very convenient to travel along, but not necessarily in the vicinity of the best railway location; the river flowing between or in direction of the termini may have precipitous, treacherous banks, be crooked in alignment, and afford not nearly so feasible a route as the upland country adjacent; just beyond a certain forbidding range of hills may lie a direct and cheap route, and a pass through the barrier may really exist, being hid in the distance by an overlap. In fact, the frame of mind suitable for such an undertaking should be optimistic, ready to believe that if only time enough is available, the best route can be found, but at each moment doubting that such a route is yet discovered.

In addition to those general economic considerations which have been touched on in previous chapters, it is well to remember, amongst other things,

(*a*) That lines following large streams will usually require heavy bridge work and masonry in crossing tributaries.

(*b*) That one bank of a river may be much better than the other, and that it may even pay to cross the river at rare intervals to secure alternately favorable stretches of construction.

(*c*) That lines on side hills are more costly to maintain than those through level country, owing to the sliding and washing that takes place.

(*d*) On the other hand, that a cross-country line, usually, will cross many summits, and even when skilfully located, and often at a considerable loss in distance, will abound in curvature and maximum grades.

(*e*) That in each locality will be met men who have an intimate knowledge of the minutiæ of the surrounding country. Many of these look on themselves as born locating engineers, and while their ideas on grades and curves are usually misty, every shrewd engineer will not be averse to the valuable aid which such men voluntarily offer; the only difficulty lies in sifting the wheat from the chaff without giving personal offense.

(*f*) That the engineer of reconnaissance and afterwards of surveys is the first officer of the railway company to be thrown in contact with the people who are to become the future patrons of the road, and, as such, his manifest duty is to make as many friends for his company as he can, consistently with his other duties, and enlist their sympathies in its favor; in this way a much more reasonable spirit will be created which will display itself when right-of-way questions begin to arise.

After a complete study of the intervening country has taken place, a rough sketch map should be made from the notes taken, and other existing ones, on which will be shown the positions of all streams, summits, etc., with elevations marked at critical points, then possible routes will be indicated, calculations made of the length of lines, maximum grades, probable amounts of curvature, approximate cost of constructions, present and future traffics, etc., all of which, although much in error, will usually narrow down the question to two or three routes which are selected as the most likely and suitable ones for instrumental surveys.

ARTICLE 13.—PRELIMINARY OR TRIAL LINE SURVEYS.

The roughest class of preliminary survey may be an amplification of reconnaissance, in which a small party of three or four men pass rapidly over several proposed routes at a rate of five to fifteen miles per day to determine what grades can be obtained before more accurate survey begins. In open country rapid progress can be made, using stadia for distances and using vertical angles for elevation or depression, which are checked by an aneroid barometer. In a wooded country the distances will be determined more rapidly by chain and compass, and heights by aneroid. Side slopes may be noted at difficult spots by some form of clinometer. What is usually wanted is to know what grades can be obtained at certain critical points, in order to adopt a ruling grade for the route. The instruments required are a light transit with stadia hairs, compass and vertical arc, a stadia rod, an aneroid barometer, a clinometer, a 100 ft. steel chain and 50 ft. linen tape. On this class of work the error of stadia measurement should not be more than 1 in 1000, which is more accurate than rough chaining. When a full survey party for instrumental work is to be equipped, a variety of causes tend to determine the men and instruments required.

(*a*) In an open rolling country. If contour lines are not needed, the party will usually consist of—

Chief of party, Transitman, Leveller.	Engineers, preferably all experienced.
Rodman, Front Picketman, 2 Chainmen.	Active young men, preferably educated college graduates, not afraid of work.
2 Axemen, 1 Stakeman.	Seasoned workingmen, used to bush life, axes, and hard work.

If under canvas, add one cook and one assistant cook, and in this kind of country always use a transit.

(*b*) In thickly wooded country, without iron ore, better results, for the same labor, will be obtained by using a 12-inch to 16-inch compass, instead of a transit, avoiding many detentions, useless cutting of trees, etc. The compass has no cumulative error, and will give good results where no contours are taken ; if contours are to be taken, it is better to establish a transit line for future use. In a wooded country two or three extra axemen will be needed to make rapid headway ; the front picketman also, in this case, should be an expert axeman, and lead the others.

(*c*) If the country is much on side-hill, another party is needed in addition to the transit and level parties, whose duty it is to take contours. In the past contouring has often been omitted, and although there have been some men of great natural talent and long experience who have been able to locate well, even through very rough side-hill country, by eye alone, yet even to such men a properly conducted contour survey would have been of great advantage. It is becoming more fully realized every day that a contour map, with a location line laid on it in the office and revised afterwards, where necessary, in the field, is a very valuable part of preliminary surveys in such a kind of country. This topography party consists of two or three men, equipped with a level board, level rod and hand level, or else with a clinometer and tape to measure side slopes ; the work is carried on one day behind the level party, and the method of procedure is somewhat thus :

Detached sheets of paper about 18 inches by 24 inches, have plotted on them the centre line and level height at each 100 feet and hub, according to the previous day's records ; these sheets are mounted on a drawing board and taken into the field, where 5 feet or 10 feet contours are plotted and sketched direct, for a distance of 20 to 50 feet in elevation, up and down hill from the centre line, depending on evident requirements ; with a little practice, the distance to each contour can be taken and plotted very rapidly, obviating the necessity of notes. Intermediate irregularities, etc., can be also sketched in by eye, and the sheets when taken back to the office can be placed in proper alignment and chainage, and a tracing taken if necessary ; but probably the projected location line will be placed on these sheets and then transferred to the field at once, or by another party following, or the whole matter may be held over until a decision is arrived at as to the correct location route to adopt; this will evidently vary with each case. If the contour notes are recorded in books in the field, they may be plotted on a continuous roll in the office; but such a method is more tedious, and little irregularities which would be sketched in the field are often omitted in notes. A topography party relieves the transitman of all note-taking except centre alignment, whereas all notes of natural and artificial topography are taken by the transitman where no topography party is employed, thereby delaying the progress of the whole survey. A topographer should preferably be a Provincial Land Surveyor also, so that his work in recording land lines and making plans may at once be legalized.

The qualifications and duties of the members of a survey party are somewhat as follows:

The Chief of Party should be a man of vigorous mental and physical attainments, familiar with the details of survey life and minutiæ, with a wide experience of construction, and even, if possible, of maintenance of railways, well informed on such matters as have been touched on in previous chapters, and capable of commanding prompt obedience and zealous assistance on the part of every member of the party. If, in addition, a man can be found who has also a natural genius for railway location, he cannot be too highly treasured or paid. The chief of a survey party is the most important officer in the pay of a railway company where location is of a difficult and perplexing nature. Crippled constitutions and receiverships are more often the result of poor location than from any other cause, hence the high value of the men who decide on such matters. A chief may be a strict disciplinarian and still command the regard of his assistants; he should have free scope to dismiss anyone not competent and willing to do good work; and should never do any work for subordinates, except in the rarest instance, but should be well on at the front most of the time, devising the next step before it is needed, and having in view a general plan of the country, not looking straight ahead, but feeling that just "beyond" there may be a better line. The rate of progress is fixed by those at the front, the others must keep up. A chief of party carries usually a pocket note book, or even topography book, an aneroid barometer and a pocket compass.

The Transitman should be an engineer of some experience, particularly in handling men, keeping full and accurate notes, and rapid and yet delicate handling of his transit. He should be alive to the general movement of the men in his party, which means that he should not always be looking through his telescope at them, but commanding their movements directly also, and above all, he should put his transit in position quickly, and not keep a whole party waiting while he dawdles over his levelling screws, etc. Where there is no topographer, the transitman, in addition to keeping notes of the survey alignment, must sketch neatly, with necessary measurements, all buildings, roads, farm lines, etc., in fact all artificial and natural topography, and obtain all owners' and tenants' names. In a level country, topography should extend for at least 500 to 1,000 feet on each side of the line, as the location may be moved that much, and thereby run through houses and barns that have not been noted. This should be done where necessary by accurate chainage offsets. In country of steep side inclinations this is not necessary; judgment will determine the width of the topography belt needed in each instance.

The Leveller may be a young engineer of limited experience, although preferably one capable of rising rapidly to higher positions, and not one whose engineering horizon is bounded by such work. In addition to centre line levels, taken at each 100 feet station, hub, and intermediate change of vertical direction, the leveller notes the wooded and cleared portions, the class of timber, probable nature of material in cuttings and borrows; the depth, volume of flow and high water mark of all streams, and establishes bench marks, at say each half mile on preliminary surveys.

The Level Rodman and Chainmen should not only be instructed how to do their work, but day after day should be made to chain and hold their rods correctly; chains should be tested frequently. It is certain that more errors are due to poor chaining and rodding, to insecure hubs, and to slovenly work amongst subordinates generally, than to poor instrumental work, although the blame for such errors is usually laid on the latter.

A Front Picketman is invaluable and should be distinct from the chainmen; he should be an active, intelligent man, one who can select a transit site with judgment, make and drive a hub well, take centre, make and drive reference stakes, make a cross-head for back-sight, and then, after placing his picket *exactly* on line, or laying it on the ground, continue to make stakes until the transitman arrives, or better still, if so directed, he may continue to the next site and be ready by the time the transit is placed, to take hub again. In cleared country, hubs should be driven in secluded spots along fence lines, etc., wherever possible, or else in a few months all traces of line across cultivated fields will be obliterated. If hubs come, necessarily, in open places, extra ones should be put in in sheltered spots. If the line is being carried through forest, the same care will not be necessary to preserve the line, and transit sites will depend more on natural profile; in this case the front picketman should be continually taking line for clearing, and leading, and commanding the axemen, being himself also, for the time, an axeman. In general, it is best to not have a back picketman; but have the transitman place a cross-head on line within a few inches of his transit telescope just before moving forward. If a back picketman is employed, it is best to still use a cross-head and keep the man merely as a guard and handyman for occasional use in emergencies; a cross-head is very accurate and never falls asleep at inopportune moments.

During survey, people naturally resent injury to crops and premises, even when the least possible is inflicted, and polite words and sincere endeavors to minimize the loss are rightful and expedient. Many survey parties constitute themselves armies of invasion; trees are needlessly cut down, growing grain trampled on, fences torn down to make stakes, and a general tone of overruling ruthless power is prevalent, all of which is wrong and foolish.

Many life-long enemies to the road have been made in just this way, and the probable immediate consequences will be that all stakes will be torn up and thrown aside as soon as the party has passed by, and, in addition, the purchase of right of way will be made needlessly difficult and expensive ; the far-reaching consequences to a railway company of the actions of survey parties in this respect are beyond calculation.

After preliminary surveys are completed, and it is desired to obtain approximate estimates of the quantities and cost of construction for a comparison of routes, various short-cuts are used. Excavation tables can be purchased or made for taking out quantities of earthwork. General plans of trestles, culverts, etc., can be used, and tables drawn up of the cost per lineal foot for various sizes and heights, but the larger structures will each require special calculations. In taking out approximate quantities, remember—

(*a*) That embankments require more than cross-section measurements indicate, by about 5 per cent. if of sand, 10 to 12 per cent. if of clay, and 15 to 25 per cent. if of loam or peat, but that rock expands 25 to 75 per cent., depending on the size of the rocks. This shrinkage will not take place fully for a year or two, and may not exist at all during hurried construction, and will be made up afterwards by train.

(*b*) That unless the depths of foundations are known, a liberal allowance should be made for possible deep ones.

(*c*) That side-hill quantities are not indicated by the centre line profile and should be specially provided for.

(*d*) That the classification of material is likely to be higher than surface indications would seem to warrant.

ARTICLE 14.—LOCATION SURVEYS.

The duties of each engineer of a survey party are considerably increased when the location of the selected preliminary line is decided on. The chief and transitman note the foundations for all structures, and should be given time and facilities to have soundings of the beds of the streams, etc., made, so as to determine the depth of foundations quite accurately. This may be made a part of the leveller's duties, if he has more time for it. They should carefully note the natural resources of the country passed through, whether good quarries and ballast pits exist or not; whether timber suitable for trestles, piles, or ties is available, and the probable traffic on various intersecting highways, so as to determine at each one whether it will justify the company in expenditure sufficient to pass the road over or under the railway. In fact every item of information necessary for a complete knowledge of each structure, etc, so as to be able to say definitely what structure is best at each point and why. If the topographer is not a land surveyor it will now be necessary to add one to the party, whose duty it will be to fix the exact position and angle of crossing of each property line, and measurements to nearest monuments, the bearings being taken by compass. Also to obtain the full names of all owners and tenants whose properties are to be affected, the exact positions of all buildings or highways within 400 or 500 feet of the centre line, and all prominent natural topography. The leveller, in addition to those duties already noted for preliminary work, will need to establish and witness bench marks about every 1,000 feet, not closer than 30 feet or 40 feet to the centre line, nor more than 50 feet or 55 feet in a bush country, while in an open country any convenient distance, laterally, will answer. Very few places should be deemed suitable for a bench mark; the root of a large green stump, or the top of a small one, cut off for the purpose, is the best. If the root of a tree is used it will grow a little in the course of years; it is liable to be invalidated by the wind blowing the tree over, or lifting the roots in loose soils, and worst of all, if on the right of way it is liable to be burnt up or cut down below the blaze made on the side of the trunk for reference, thereby being lost entirely. Bench marks should be selected at elevations close to structures, but otherwise at heights nearly that of the proposed grade line, so as to be convenient in running grades, ballast heights, etc. When a located line is laid down without much revision, it will, in any difficult country, pay by many times the expenses of a survey party, to revise the whole line, when numerous small changes will be made from point to point.

In finally staking such a revised line, it should be thoroughly done, and hubs and stakes stout and well driven, good strong nails (1-inch tinned) used. No hub should project more than half an inch above the ground, and in a settled country stakes should be bought at a saw-mill and carried along a day at a time; the using of fence rails for stake material creates more ill-feeling than the trifling economy warrants. Unless construction is imminent, all hubs should be referenced by cross lines or otherwise. Nothing is much more disheartening to the constructing engineer than to find a located line almost

Fig. 19.

Methods of producing a straight line past an obstacle

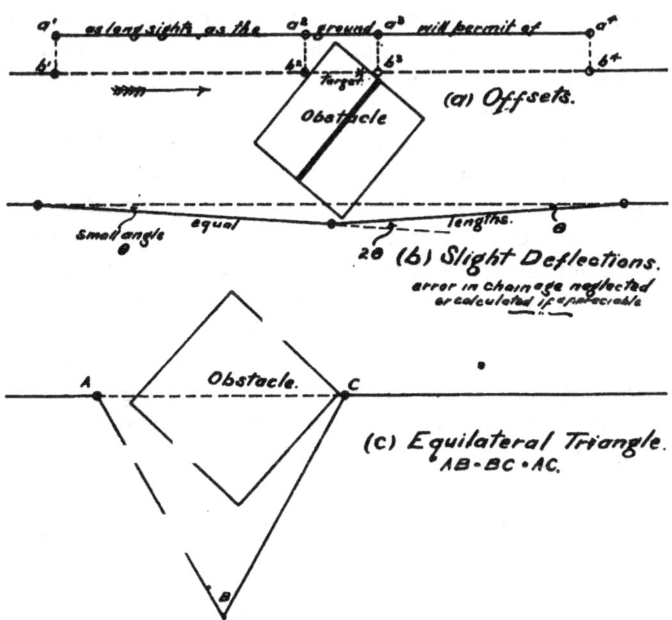

obliterated and untraceable. There are three general methods of prolonging a straight line beyond an obstacle.

(*a*) By offsets, where the necessary offset is not very long; this is the most accurate method. The measurements a_1b_1, a_2b_2, a_3b_3, a_4b_4 are identical and made very carefully with a steel tape and plumb line; the transit sites would be at b^1, a^3, b^4, with a target placed on top of the obstacle, if possible, as a back sight check. (See fig. 19.)

(*b*) By making a slight angular deflection measuring a certain distance until just opposite the obstacle, then deflecting back twice the first deflection, measuring an equal distance and then deflecting again on to tangent, by an angle equal in amount and direction to the first one, the error in chainage is usually disregarded; this method introduces three angular measurements and is not likely to give an exactly straight line for this reason. (See fig. 19.)

(*c*) By laying out an equilateral triangle, this fixes the chainage beyond the obstacle, and presumably the direction and line; but as this method introduces three angular measurements and two linear ones, it is not apt to give as good results as the first two. (See fig. 19.) It is understood that if by placing a transit on top of a secure obstacle the line can be prolonged directly over it, it is best to do so even at considerable personal inconvenience.

If very accurate transit work is desired, it is not best to trust to the adjustment of the instrument, but take two points on each hub, and use the mean. In the same way, equal backsights and foresights in levelling should be obtained wherever possible, to minimize the result of a level being out of adjustment, and also it is best to adjust instruments for about the distance that the ordinary sights are to be in any given class of country. The travel of the tube in a large change of focus often throws an instrument out of adjustment for very short or very long sights.

It is often found that a survey party, before being disbanded, has time to do cross-sectioning for construction; this is a mistaken economy, and a source of errors and mutual accusations. The members of a survey party do not take interest in work they are not to superintend, and the cross-sections will probably be poorly chosen and executed. Then the centre line will very likely be altered in various places, which will invalidate all sections at those points. Generally speaking, it is best to have the engineer of construction do everything of an engineering nature which appertains directly to his work.

CHAPTER V.

ROADBED CONSTRUCTION.

ARTICLE 15.—WATERWAYS.

The construction engineer, after retracing the centre line, and checking levels, and establishing additional B. M.'s, if necessary, should verify and complete the list of structures fixed upon by the survey party.

The class of structure will depend upon the money and material available, but its cross-section, if it is a waterway, will depend on the *maximum* flow of water it is expected to carry, while if it is a cattle pass or public crossing, its *minimum* dimensions will be fixed by law. Many causes affect the maximum flow of water across a railway roadbed, at a given point, besides the drainage area; in the case of small streams or local watersheds, the building of the roadbed, and consequent roadbed and catch-water ditching, will concentrate the flow, from quite a large area, on a culvert that would naturally have had much less flow to accommodate; this should be anticipated. Then, again, the construction of a railway in a new country will induce such activity as will cause large tracts of forests to be cleared off, and in a few years these cultivated areas will allow storm waters to pass off more rapidly than when the same area was in forest, which should therefore be anticipated and provided for. If the drainage area is in a nearly level country, water will arrive at a given point more gradually than if the slope of the country is abrupt; and also the shape of the drainage area and distribution of tributaries has a marked effect on the maximum flow. If a long stream has few and small branches, the maximum flow will be much less than though there were more and larger tributaries and less main stream, the total area being the same, especially if the branches empty just above the railway. In this case the flood water from all of them may arrive about the same time. Stony ground, also, sheds water much more rapidly than mellow and highly cultivated ground, and small areas are more liable to *abnormal* floods than large ones, because cloud bursts seldom occupy large tracts of country.

All such considerations should be weighed along with that of acreage, which should be determined, roughly, by personal examination, for every area large or small draining toward and across the railway under construction. There are several empyrical formulæ, purporting to connect the square feet of waterway required with the acreage drained, but they, necessarily, contain a co-efficient which varies with so many causes, such as those just given, as to make them difficult of application, even leaving out of question, the variation in rainfall in different localities. Indeed, it is the greatest rainfall for short periods that is the most important factor, and records of this are usually deficient.

The safe carrying capacity of a box or arch culvert may be made a maximum by digging straight wide approaches and offtake ditches, and by building flaring wings at each end to avoid contraction, and may be abnormally increased by designing it to carry a head of four or five feet of water in an emergency, which of course, increases the velocity—this, however, is hardly safe practice.

Baker's "Masonry Construction" has these formulæ:

(1) *Myer's.* — Area of waterway in square feet = $C \sqrt[2]{\text{drainage area in acres}}$. In which $C = 1$ for rolling prairie, $1\frac{1}{2}$ for hilly ground, 4 for rocky precipitous ground. This formula, Baker considers, will give too large results for small areas, and too small results for large ones.

(2) *Talbot's.* — Area of waterway in square feet = $C \sqrt[4]{(\text{Drainage area in acres})^3}$. In which, $C = \frac{2}{3}$ to 1 for rocky precipitous ground, $\frac{1}{3}$ for rolling ground, having floods and snow at the same time, and $\frac{1}{5}$ to $\frac{1}{6}$ for long narrow valleys with little or no snow. This formula, used with judgment, will probably give as good results as can be expected, where there are so many varying conditions.

Aside from any data as to acreage, etc., the high water mark at some narrow point in the channel may be noted, information from old residents as to abnormal freshets gathered, the waterway under any existing highway bridges measured, and any other influences noted bearing on the maximum flow, such as the rain records for past years for the nearest weather station, and the probability of the maximum flow being increased by clearing the country, if at present in forest, etc.

It is best to err on the large side, although some engineer has said that if a road has *no* wash-outs from too small waterways now and then, the structures are too large for ultimate economy.

ARTICLE 16.—STRUCTURES FOR SHALLOW EMBANKMENTS.

The first consideration is to get water across the roadbed and away from it as quickly as possible, this is an axiom of good drainage. To do this where the embankment is from 6 inches to 2 feet deep, and the drainage area only nominal, is often a puzzle. We do not wish to leave an opening in the track, and pipes or stonework are impossible; the usual course is to fill in the pocket above the bank and drain through the track by track boxes (Plate 1, Fig. 20), or where there is from 18 inches to 3 feet of a bank in place of an open culvert which some engineers put in, a plank box (Plate 1, Fig. 21) is preferable, as it can be replaced easily when rotten, will stand the vibration of trains, and does not leave any opening in the track. Sometimes with a very slight drainage area, a blind or French drain is used, which consists of small flat stones placed so as to give a triangular opening of, say 6 inches high x 8 inches wide, but such a waterway is liable to get choked up with leaves, etc., and cannot be depended on indefinitely.

When we come to banks of from 3 feet to 6 feet in depth it is usual to employ iron, terra-cotta or concrete pipes if the waterway is small, and where a heavy flow of water necessitates it, open culverts of from 4 feet to 8 feet span. These latter may be of timber, stone, concrete or brick, but should always be surmounted by an ordinary trestle floor. The use of stringers only is an abomination and a death trap, which should not be tolerated.

Culvert Pipes.—The use of double-strength, well-burnt sewer pipe for culverts has increased rapidly of late years in certain sections. They fill a certain want, either where stone is scarce or absent, and at points distant from water or rail communication, as the cost of teaming is small compared with that of iron pipes. In a country subject to severe frost certain precautions must be taken to avoid water settling under or in the pipes, which will crack them in freezing. The grade must be ample, care taken to have the grade convex longitudinally of the pipe rather than concave, the joints made watertight with cement mortar, and headwalls built at each end with deep aprons, to avoid having any water flow along the pipe outside of it. In southern localities of the United States of America, pipes are used more freely and carelessly, head walls and joint filling are omitted, or only a timber headwall used, but such omissions will invite trouble in Canada.

Again, if the bank is shallow, the same care is hardly necessary in ramming back the filling around the sides of the pipe before loose dumping is commenced as if it is a deep one, but in either case care should always be taken to cut a concave bed for the pipe to rest in and grooves for the spigot joints, or otherwise the load will all come on isolated spots, and the pipe will tend to crack into four segments, bulging out at the sides and down at the crown. If the bottom is solid rock or bouldery, the condition is worse, and breakage can only be avoided by filling in some soft clay well rammed to bed the pipes in, or, better still, to bed the pipe half-way in cheap concrete, as shown on Plate I., Fig. 22. This figure also shows design for a headwall and paving at the lower end. The spigot ends of the sections are always laid up grade. As with other more important structures, pipes should be laid at such a depth that the outlet ditch leads the water to a safe distance with a gentle grade, so as to prevent undermining the lower end.

In place of sewer pipes of clay, there have been isoated attempts at using concrete pipes, but only sporadically. The choice would be entirely a matter of cost. On the other hand, the use of cast-iron culvert pipes is quite common. They can be made up to six or eight feet in diameter, the lengths decreasing as the diameter increases, so as to keep down the weight of a segment. If carefully coated with tar mixture rusting is very slow; and although such pipes are not used often during constructi n, owing to their great weight, which is against them in hauling by teams, they have a special function which is for use, when their transportation can be by train, in replacing wooden box culverts by drawing through

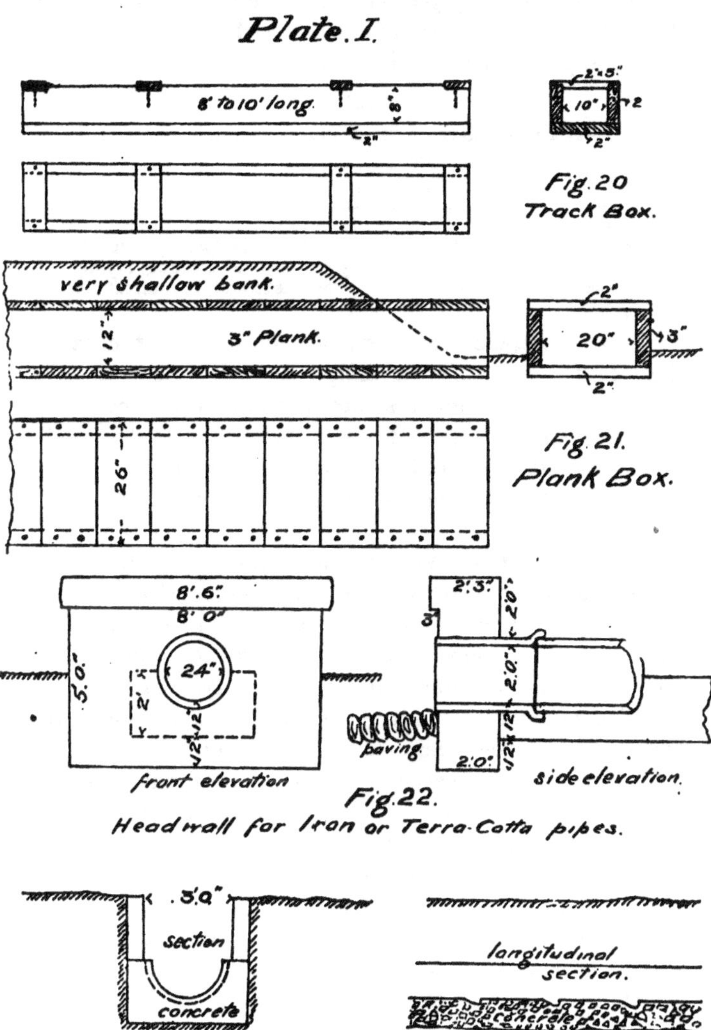

when the wooden ones are about decayed, and in case this has been anticipated· the wooden box culverts will have been made larger than necessaıy, sufficient to allow of this being done Cast iron pipes will be laid in the same manner as sewer pipes, except that the joints should in this case be caulked and leaded as with water pipes, although sometimes this is omitted; the cost per foot for cast iron and sewer pipes at the nearest railway depot to the structure will vary somewhat with the locality, but will be approximately as follows:

TABLE XII.

APPROXIMATE COST OF PIPES (NOT INCLUDING LAYING.)

Diameter.	Double Strength Sewer Pipe. (Cost per foot.)	Cast-iron Pipe. (Cost per foot.)	
12-inch	35cts.	$1 15	This does not include cost of hauling, laying, headwalls or foundations.
18-inch	70 "	1 90	
2 feet	$1 30	3 00	
3 feet		4 80	
4 feet		8 00	

The difference of cost, as shown in the table, the less cost of handling and laying of the sewer pipe and its absolute freedom from corrosion will always be greatly in favor of its adoption where well-burnt, salt-glazed, double-strength sewer pipes can be obtained within reasonable length of railroad and team-haul.

Open Culverts.—Where a large flow of water is to be carried across a shallow bank some engineers use two or three lines of pipes, but the danger of this method lies in the possibility of debris collecting around the middle walls and gradually choking up the waterway. This can be guarded against by building a screen or paling some distance above the entrance, which catches the debris. Generally speaking, however, large streams and shallow banks demand open culverts. In many cases these may answer the double purpose of waterway and cattle guard, or waterway and cattle-pass, for giving passage for cattle under the track. Such structures may be of timber, stone, or concrete or brick walls, capped with stone; but whatever kind may be used, they should be decked with a complete trestle floor, such as to make them safe for derailed trains to pass over. And indeed, latterly, some roads are adopting a solid timber floor, on which the ordinary road ballast is laid, or better still, a floor of discarded steel rails, laid longitudinally, filled in with concrete and covered with ballast; in either case the roadbed is continuous, and free from danger by derailment or fire, and presents a nore elastic and uniform bearing for the track ties.

On Plate II. (Figs. 23 and 24) are shown plans for a 6-foot open culvert of timber or stone. If the bank were deeper, the stone walls would need to be thicker, being designed as level retaining walls, and the timber culvert would need a more thorough system of interior struts, etc., for stability. If the embankment cross-section were to show a rapid descent just at the mouth of the culvert, it would be more economical to place the stepped wings (Fig. 24) at right angles to the walls, in the form of head walls, about six feet from the centre line. This is not done, ordinarily, because less economical, less stable, and subject to vibration and thrust from the train.

The timber open culverts should be well drift-bolted in each course, and have the stringers also notched down slightly and drift-bolted to the walls—the mud sills well sunk into the solid earth, and preferably with paving between them and a sheet piling apron at each end to prevent under-flow and undermining, as shown on Plate IV. (Fig. 28.) If the foundations are not good, a structure, on piles, similar to the one shown on Plate III. (Fig. 25), will need to be used. The earth being retained by a layer of four inch to six inch cedar flatted on three sides, and the two walls held vertical by drift-bolting and notching down the stringers, or if necessary, by additional struts placed from top to top of piles as shown in the figure. The use of high framed timber openings on mud-sills, lagged behind with cedar like that in Fig. 25 is not advised, they are not stable and are liable to be undermined. Wherever a depth sufficient for a cattle pass or farmers' undercrossing is required, it is better to put the structure on well driven piles extending up to grade, if a stone opening cannot be afforded.

The valid objections to open culverts with vertical walls are :

(*a*) That the structure being fixed in elevation, offers a rigid support to the track which, on banks, and on freshly made ones particularly, is elastic and settles down for several years, and rises and falls with the frost ; therefore, at such structures there is more or less of a hump, and always a poor piece of track.

(*b*) That in case of the timber culverts, the lagging behind the piles rots quickly, and is rather awkward to replace.

These considerations have led to the use, especially in the southern United States, of a form of structure shown on Plate III. (Fig. 26), which consists of two bents of piles,

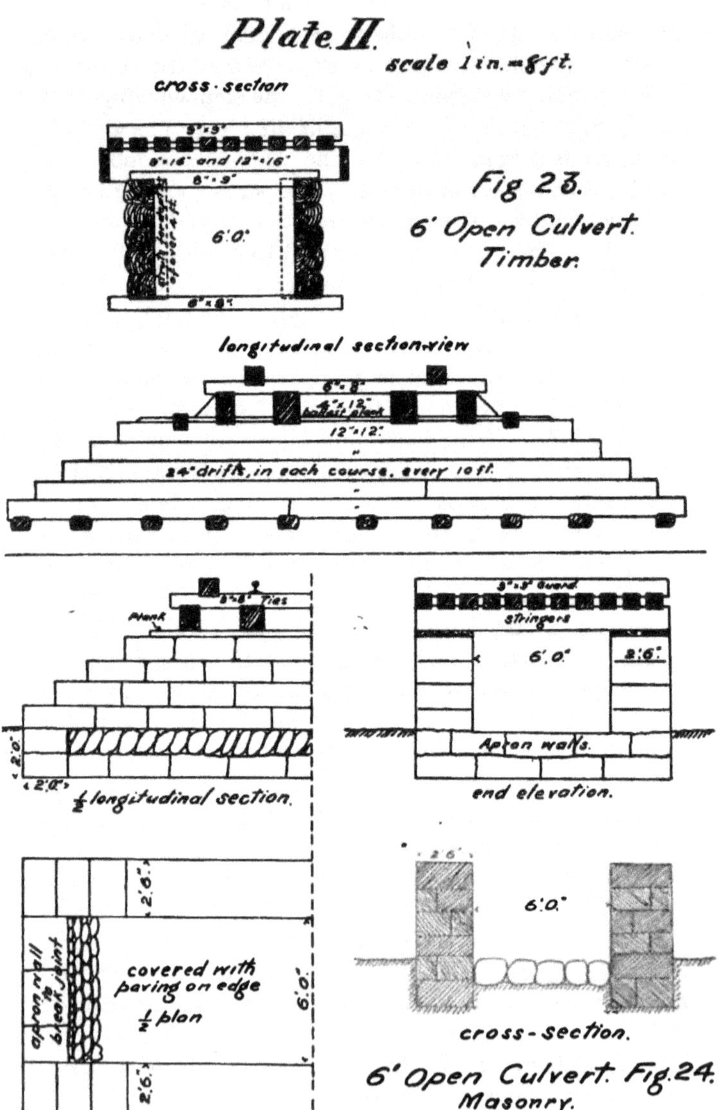

Fig 23. 6' Open Culvert. Timber.

6' Open Culvert. Fig. 24. Masonry.

or two frame bents on pile foundations, with three 15 foot spans of trestle floor, having the two end supports made of mudsills resting well on to the banks. It is probable that 45 feet of trestle floor is not appreciably more dangerous than 15 feet of it, and the only valid objection to this form of structure is that the frost will heave the ends out of surface in climates like that of Canada or the northern United States, but its openness for repairs, the elasticity of the ends which rise and fall with the embankments, its freedom from rot, except the easily replaced mudsills, and the possibility of an enlarged waterway by rip-rapping the sloping banks, to allow for an exceptional flood, are all points much in favor of such a structure. This structure is evidently limited to banks less than eight feet high.

The class of masonry for open culvert walls will need to be superior, owing to the effects of vibration from the trains, to avoid part of which oak planks should be placed under the ends of the stringers. The class usually specified is second-class bridge masonry, and will cost from $8 to $10 per cubic yard upward, depending on the quantities in each structure and total quantity in the contract. The economy of rubble concrete walls capped with a stone coping is being now recognised.

The cost of structures of these styles will be approximately as given in Table XIII., taking masonry at $10 per cubic yard, including foundations; paving at $3 per cubic yard; sawn timber at $30 per M.B.M., in place including iron; cedar lagging and timber walls at $25 per M.B.M. in place, and piling at 30 cents per lineal foot, driven, say, 10 feet into the ground.

TABLE XIII.

APPROXIMATE COST OF OPEN CULVERTS.

Structure.	Height of Waterway.	Clear span in feet being 6 feet.	8 feet.	10 feet.	12 feet.	15 feet.
		$	$	$	$	$
Timber opening	4 feet.	106	116	126	135	148
Timber walls.	6 "	157	167	177	186	201
(Fig. 23.)	8 "	212	222	233	243	259
Timber opening,	4 feet.	110	118	125	133	145
piles and lagging	6 "	144	152	158	167	179
(Fig. 25.)	8 "	168	176	183	191	203
Three span.	4 feet.			170	203	251
Opening on piles.	6 "				208	256
(Fig. 26.)	8 "					267
Stone opening.	4 feet.	317	341	365	389	426
Trestle floor.	6 "	457	481	505	529	566
(Fig. 24.)	8 "	608	632	656	680	717

From which it is evident that piles with lagging is slightly the cheapest, except with the smallest height and span, and that at 8 feet high and 15 feet span the three-span opening comes to about the same as the other timber structures. The cost of the stone opening is from two to three times as great as the timber ones in first cost, at $10 per cubic yard, but in many cases this could be materially reduced by using concrete at $6 to $8 per cubic yard, at which price a very superior quality can be made even in small quantities. An interesting feature of this table is the deduction that the length of span affects the cost so slightly, it will hardly pay to risk anything in size of waterway for such trifling economies.

ARTICLE 17.—SMALL WATERWAYS WITH HEAVY EMBANKMENTS.

Under these conditions pipes may still be used, if care is taken in laying them; up to any height, if the waterway is very small; but for cross-section areas of four square feet to twenty square feet, the structure commonly used is the box culvert, which may be made of timber, stone, concrete or brick. The two latter, however, being used, usually, in the arch form, as otherwise stone covers are necessary.

Timber Box Culverts.—These are used where cheap structures are desired, or often in undeveloped districts where construction is hurried, timber plentiful, and stone scarce, they should not be put under embankments more than 12 feet to 15 feet high, unless built large enough to admit iron pipes that will carry the rainfall after the timber culvert has begun to decay, which will be in six to twelve years, depending on the timber, etc. If the bank is a shallow one, it will not be very expensive to replace the decayed timber culvert by another similar one, or by a stone box culvert, at a time when stone can be cheaply delivered by rail and the company can afford the outlay, and if the covers are made long, as in Plate IV. (Fig. 27), they will hold up for a year or so after the side timbers have started to rot. Of the two styles shown, the one (Fig. 28) is superior in some respects. It is fastened by iron drift bolts, instead of oak tree nails. It has a row of sheet piling driven at the ends to prevent underflow and undermining, and has solid paving laid between the mud-sills, all of which are distinct improvements. For such structures, probably, cedar is the most durable wood, and pine next. A distinct advantage of timber box culverts is that on soft swampy foundations, all that is necessary

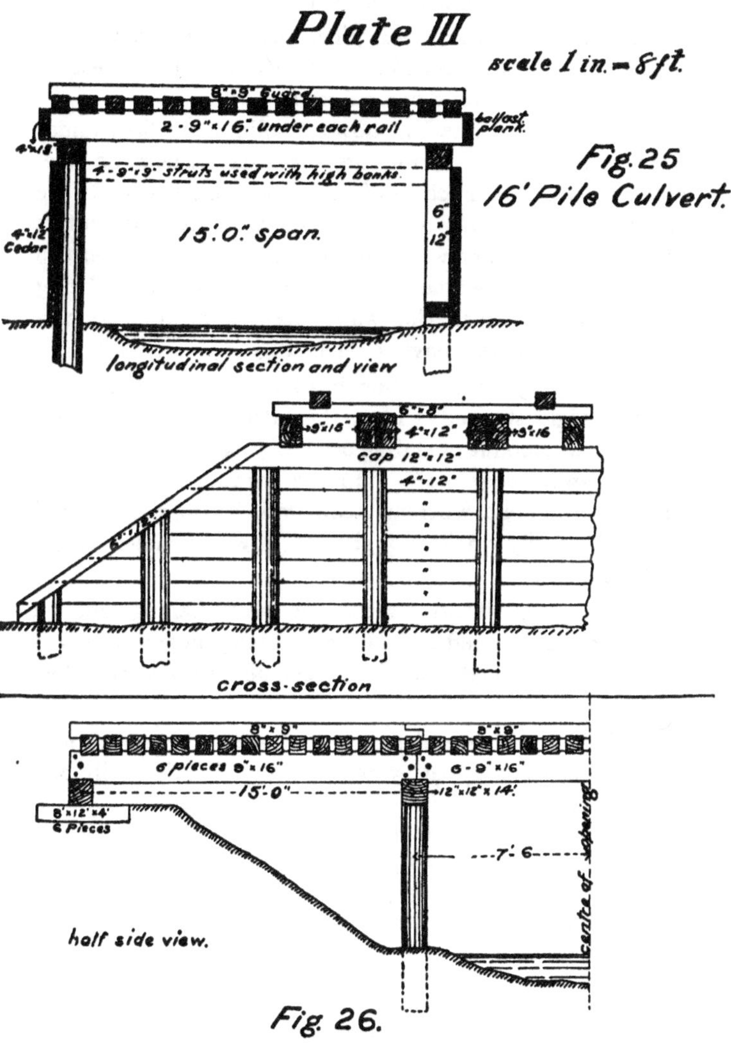

Plate III

Fig. 25
16' Pile Culvert.

Fig. 26.
Three-span Opening.
(in place of one 15' span pile culvert)

is to make a wide solid floor of timber instead of mud-sills, and even lay this floor on several sills running length-wise of the culvert to distribute the loads over the whole area. Even though some settlement should occur, the elasticity of the timber will save the structure from damage, whereas with a stone or brick culvert any serious settlement means destruction to the masonry.

Stone Box Culverts.—Typical plans are given (Figs. 29, 30, 31, 32) on Plates IV. V. and VI. to illustrate essential differences in stone box culverts.

(*a*) Fig. 30, Plate V., shows a solid stone floor under walls and for paving, while the others have the walls independent, which is much preferable because the loads are carried symmetrically to the foundation, *i.e.*, the centres of pressure are in line with the centres of resistance, because the paving may become dislodged without the walls being injured and because the walls may, if desired, be carried lower than the paving, as in Fig. 29.

(*b*) Figs. 30 and 32 show head-walls while the others have straight-stepped wings, the latter is better practice because no amount of sliding or thrusting of any kind from above, can dislodge more than the parapet wall, which is only an ornament, whereas head-walls as in Figs. 30 and 32 can be easily cracked or thrown down by slides in the embankment.

(*c*) Fig. 31 illustrates the use of corbelling where a wide span is required and stone for long, heavy covers is scarce, this method may be developed into a complete gothic.

(*d*) Fig. 29 has a distinctive feature in the *well* formed by the projecting upper wings, which will effectually prevent blocking the mouth with debris, because if any collects here it will merely form a dam, over which the water will pass safely and fall into the well thus formed, whereas with the other styles shown, a complete blockade might occur; this is a matter of importance in wooded countries.

Stone culverts in cold climates are laid in cement mortar, including covers, and the paving is flushed with grout until full; and in all climates apron walls should be sunk two feet to three feet at each end, to prevent leakage along and under the walls and paving, otherwise the action of frost and undermining will both be destructive; but in mild climates such culverts are usually laid dry, and if attention is paid to the bonding and laying, the structure may last indefinitely. This class of work is shown in Fig. 32. Wherever the fall is rapid, paving

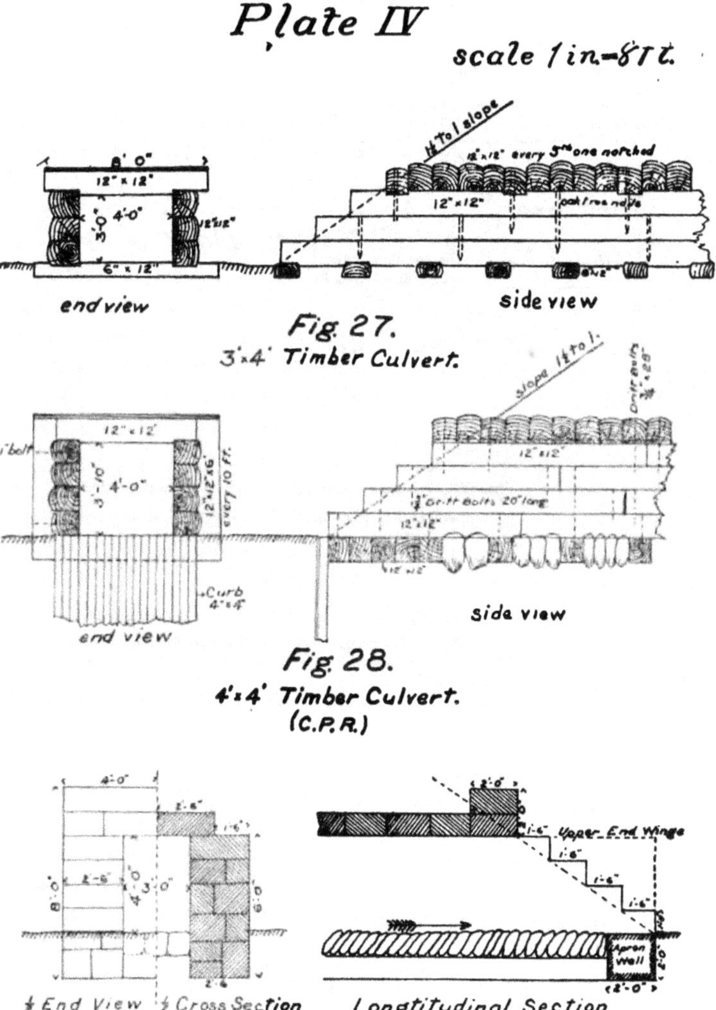

Fig. 27.
3'x4' Timber Culvert.

Fig. 28.
4'x4' Timber Culvert.
(C.P.R.)

Fig. 29.
3'x4' Masonry Box Culvert.

should be laid beyond the lower end, as in Fig. 30, and may even consist of a very heavy flat stone floor, if the grade of the culvert is excessive. Culverts should, of course, be laid to the natural cross section, no matter

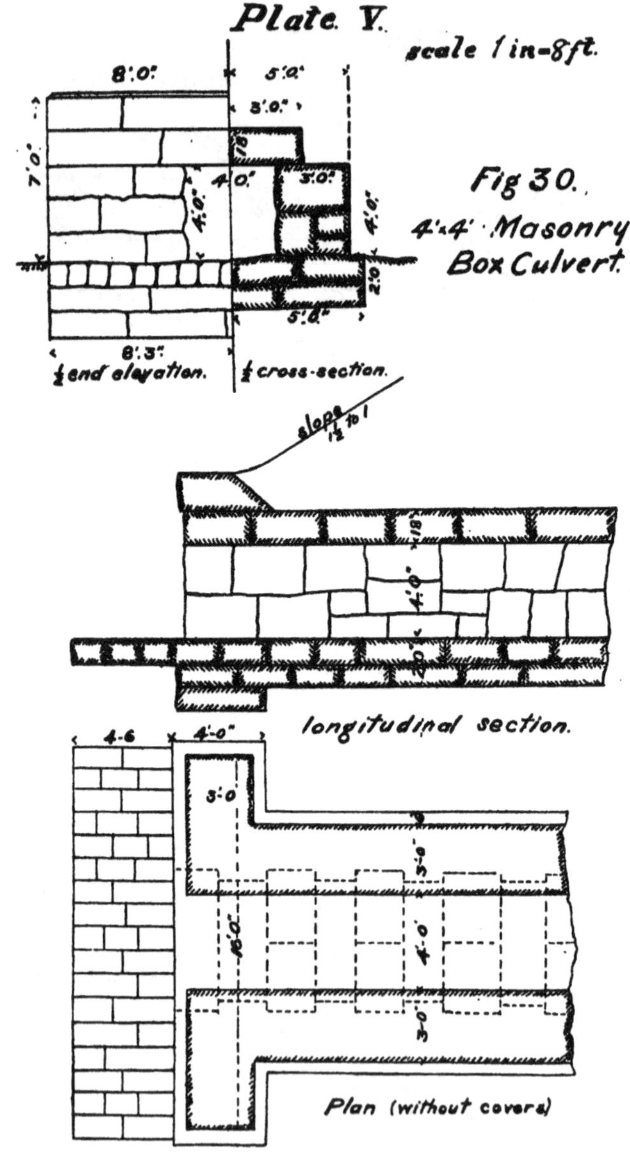

how steep—in order to get the outlet low enough to prevent undermining, the direction of the discharging stream, in plan, is immaterial—water will get away somehow, but, in profile, there should never be an increase in the rate of fall, just below the lower end, unless on solid rock. If the foundation bed is solid, the core which holds up the paving may be left of correct height to carry it, but if the foundation is poor, it will be best to build first a layer of concrete one foot to two feet thick, and commence masonry work and paving on this, or, in case the foundation is always to be under water, a grillage (platform) of timber will be suitable, as in timber box culverts.

Weak foundations are often the cause of failure in stone culverts, and all doubtful ones should be tested by the engineer himself, by driving an iron bar down in several places, and it is best to be on the safe side; a little sediment on the paving will do no harm, and will be swept out at each storm, whereas if the discharge end is too high, first a hole is worn, and finally the lower end is undermined and falls down.

If possible, culverts should be located at right angles to the centre line, and this can usually be done by diverting the entering stream, and using the material in the embankment adjacent. Skewed structures are expensive in many ways, more particularly, however, with arched culverts. The inspection of stone culverts during construction should be a rigid one, as rascally work can be hidden quicker in this class of masonry than in any other. Especially inspect the covers as to soundness and proper bearing on the walls, which should be from 9 inches on small culverts to 15 inches for large ones; they should have full bearings at each end, and be well spauled and mortared at the joints, to keep out earth and water. In bringing embankments against all culverts, care must be taken not to shove them over; filling should, if possible, be carried on on both sides at the same time, but if not, then earth should be shovelled over, up to the level of the top of the covers, before a high bank is brought forward. These remarks apply more particularly to arched culverts. The use of solid concrete box culverts to take the place of masonry ones is on the increase. They can be built cheaper, and when a knowledge of the science of cements and proper concrete making is more general, such construction will be largely adopted.

Specification for Stone Box Culverts laid in mortar.—"Culvert masonry shall be built of good, sound, large, flat-bedded stones, laid on their natural and horizontal beds. The stones used must not be less than three feet in

Plate VI

scale 1 in-8ft.

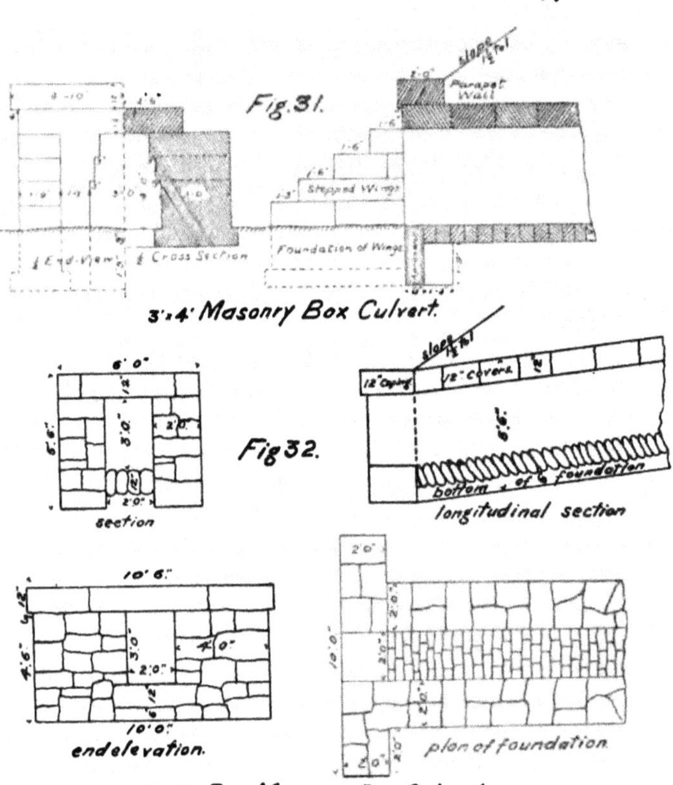

2'x3' Dry Masonry Box Culvert.

area of bed, nor less than eight inches thick, and must be hammer-dressed so as to give good beds with half-inch joints or less. Headers shall be built in the wall from front to back, alternately, at least one in every five feet of wall and frequently in the rise of the wall. The least width of bed for stretchers shall be twelve inches. In larger structures, all stones must be heavier in proportion, every attention must be paid to produce good bond, and to give the whole a strong, neat, workmanlike finish. All dimensions must be according to plans, but these may by varied if the engineer so requires."

"The paving shall be of stone set on edge, twelve inches deep, packed solid, of an even face, and inclined in direction of the stream."

"The mortar shall consist of one part good quality Portland cement to three parts of clean sharp sand, and all joints, beds and interstices shall be carefully filled with mortar and packed solid—the exterior faces and interior of barrel shall have all joints raked and pointed with mortar, consisting of one part cement to one part sand."

Cost of Box Culverts.—Taking timber in place, including iron and foundations at $25 per M.B.M., culvert masonry at $6 per cubic yard, and paving at $3 per cubic yard, including foundations. The cost of box culverts according to figures (28) timber, and (29) masonry, are given in table XIV.

TABLE XIV.

APPROXIMATE COST OF BOX CULVERTS (16 FOOT EMBANKMENTS).

Structure.	Waterway.	Total cost for depth of top of paving below subgrade.					
		10 ft.	20 ft.	30 ft.	40 ft.	50 ft.	60 ft.
		$	$	$	$	$	$
Timber Box......	2' x 3' high	183	306	429	552	675	798
Fig. 28.	3' x 3' "	213	356	500	643	787	931
"	3' x 4' "	234	397	560	723	886	1,049
"	4' x 4' "	263	446	629	812	995	1,178
"	4' x 5' "	281	483	686	888	1,091	1,294
Stone Box	2' x 3' "	254	420	587	754	920	1,086
Fig. 29.	3' x 3' "	267	444	620	797	974	1,151
"	3' x 4' "	364	607	851	1,094	1,338	1,582
"	4' x 4' "	385	645	905	1,165	1,425	1,685
	4' x 5' "	501	848	1,195	1,542	1,889	2,236

From which table it is evident that the stone culverts increase in cost much more rapidly than the timber ones, owing to the necessary increase in the thickness of the stone walls, being estimated at 2 feet, 2½ feet and 3 feet thick for culverts 3 feet, 4 feet and 5 feet high (in the clear) respectively. It does not pay, evidently, to build small timber culverts, other things being equal.

ARTICLE 18.—LARGER WATERWAYS WITH HEAVY EMBANKMENTS.

When a single box culvert 4 by 5 feet in cross section or, with very long covers and corbels, possibly 5 by 5 feet, will not carry the maximum flow of a stream, we must either use double or treble box culverts or an arch culvert. The intermediate walls of double box culverts may be made pointed to divide the flow of water, and a screen or paling may be erected some distance up stream to catch driftwood, but, even at best, their use is doubtful for the same reason as with double lines of culvert pipes, *i.e*, the danger of logs, etc., choking up the entrance; whether an arch culvert of equivalent area will be cheaper than such a structure will depend on the availability of brick, cement or cheaply-cut stone for arch sheeting on the one hand, or of large-sized stones for covers on the other.

ARTICLE 19.—ARCH CULVERTS.

The selection of materials for the construction of arch culverts will depend on circumstances; where good weathering stone can be easily quarried and cut in the vicinity it will be usually used, but if stone is scarce or costly, and well-burnt brick plentiful, then brick may be found cheaper; of course brick so soft as to be unable to stand erosion or frost should never be used on exterior faces or for the arch sheeting. The use of concrete for arch culverts is yet a very occasional one in America, but is likely to steadily increase as we have more skilled civil engineers who are familiar with the production of a cheap concrete with superior exterior finish, capable of standing frost and erosion and certain to remain sound for an indefinite number of years, which necessitates using absolutely sound, high-grade cements, and until an engineer has the opportunity of making certain of his cement by systematic testing, he is advised to avoid the use of any but the very smallest monolithic arch culverts, although, of course, their construction presents no structural difficulties, beyond the precaution of defining occasional lines of separation in the arch sheeting so as to avoid irregular contraction lines.

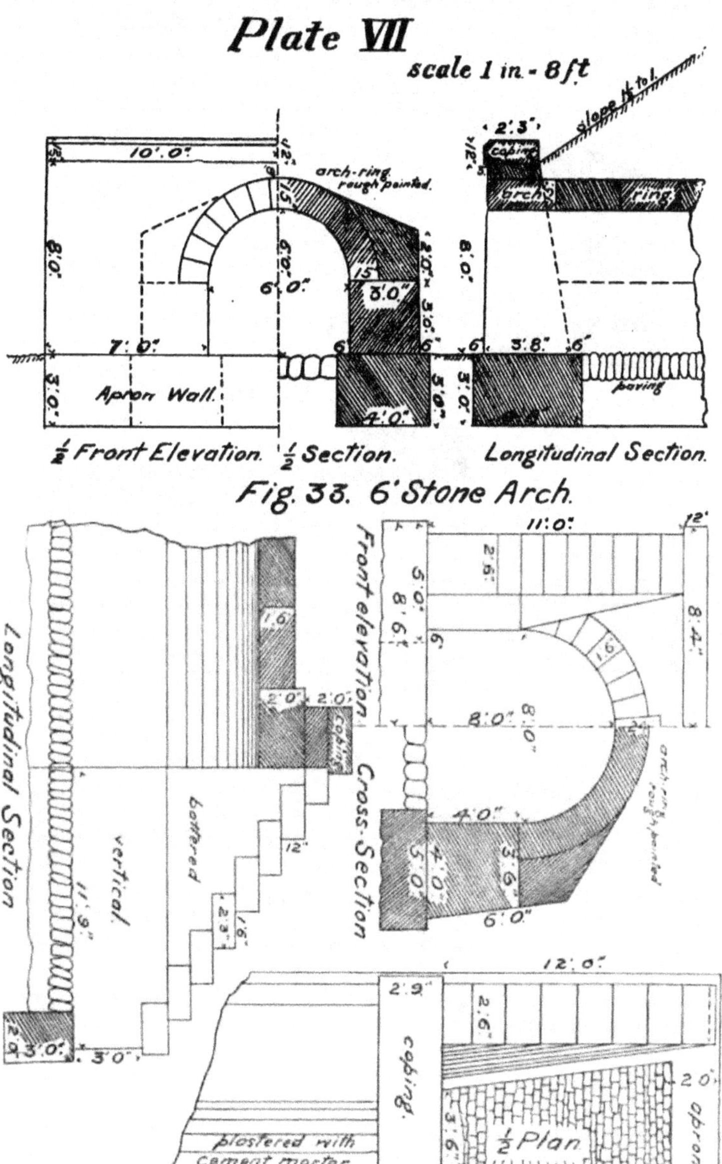

Fig. 33. 6' Stone Arch.

Fig. 34. 8' Stone Arch

Plate VIII

Scale 1 in.-12 ft.

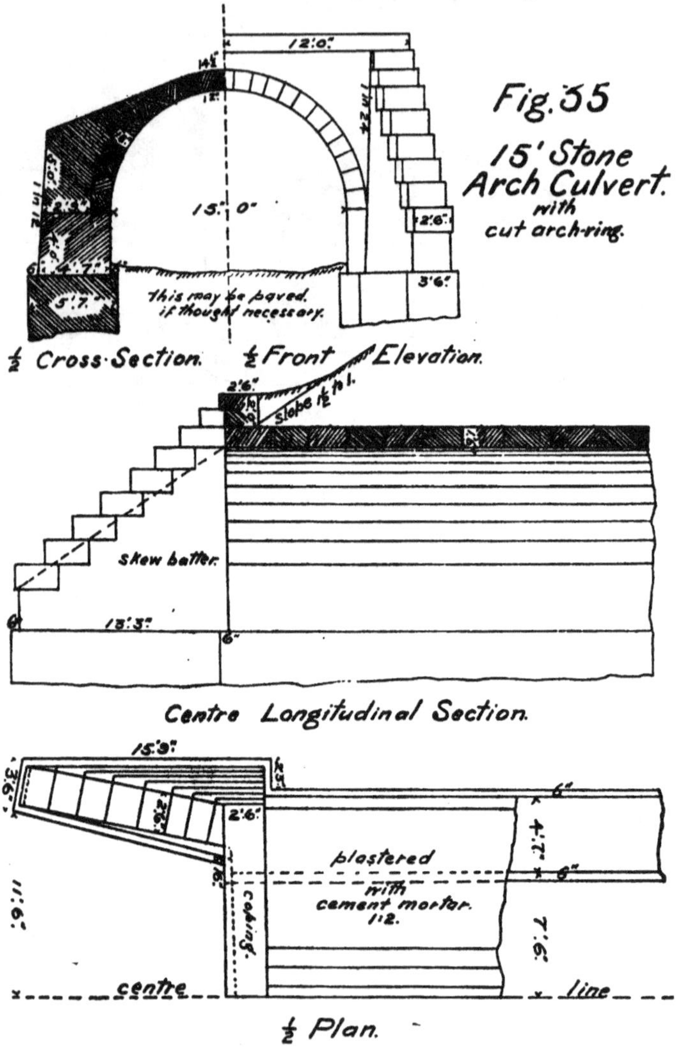

Fig. 35

15' Stone Arch Culvert with cut arch-ring.

½ Cross-Section. ½ Front Elevation.

Centre Longitudinal Section.

½ Plan.

The chief features of arch culvert designing are:—

(*a*) The shape of the end walls.

(*b*) The depth, class and form of the arch sheeting.

(*c*) The dimensions of the arch abutments.

(*a*) The shape of the end walls will depend on the span of the arch and its rise.

For small semi-circular arch culverts, say from 5 to 8 feet span, the retaining head-wall shown, Plate X., Fig. 39, and Plate VII., Fig. 33, is generally used; for segmental arches of somewhat longer span the same may be advantageous, but as soon as a larger retaining wall becomes necessary its use should be abandoned in favor of stepped wings; the reason for this is that a surcharged retaining wall, with nothing but mortar to bond it to the back of the ring stones and often loaded with wet, slippery clay filling, is liable to be displaced, unless made very heavy, and thus the designs, as shown, Plate VII., Fig. 34, and Plate VIII., Fig. 35, of the types shown on Plate X., are found more suitable. The choice between straight wings and flaring ones, or between wings flush with the faces of the barrel of the arch, and those set back clear of the ring stones will depend much on the taste of the designer; for small spans liable to catch driftwood the choice should rest on flush wings, with some flare to avoid contraction, but with larger spans, of say 15 feet or over, a wing set back so as to show the arch ring stones will have a better appearance, and give equally good or better bond between the wing and the abutment or parapet wall. The small parapet wall of a culvert with stepped wings is well buttressed and very stable; the wings themselves usually have a face batter of 1 in 12 to 1 in 24, and a section at any point suitable for a level retaining wall (*i. e.*) about $\frac{4}{10}$ height + batter, their length will be economically curtailed at a point where the steps are 2 feet or 3 feet above the ground level. Stepped wings are preferable to those with inclined copings, as the latter are liable to become dislodged in time, and do not give an easy means of climbing the bank, and, also, the coping of a parapet wall of a brick arch culvert should preferably be a stone one, as bricks are liable to be displaced by ties, boulders, etc., rolling down the bank.

(*b*) The form of the arch will depend on the depth of bank; wherever headroom permits, a semi-circular arch is used, partly because the arch sheeting stones are less expensive than those for any other than segmental arches, being all cut from one template, and partly because the abutments need not be so heavy; but as the quantity of cut arch sheeting is greatest in a semi-circular arch the saving is not very great on the structure as a whole, but when the depth of bank is the limiting feature, a much greater waterway can be obtained by the use of arches of

Plate IX

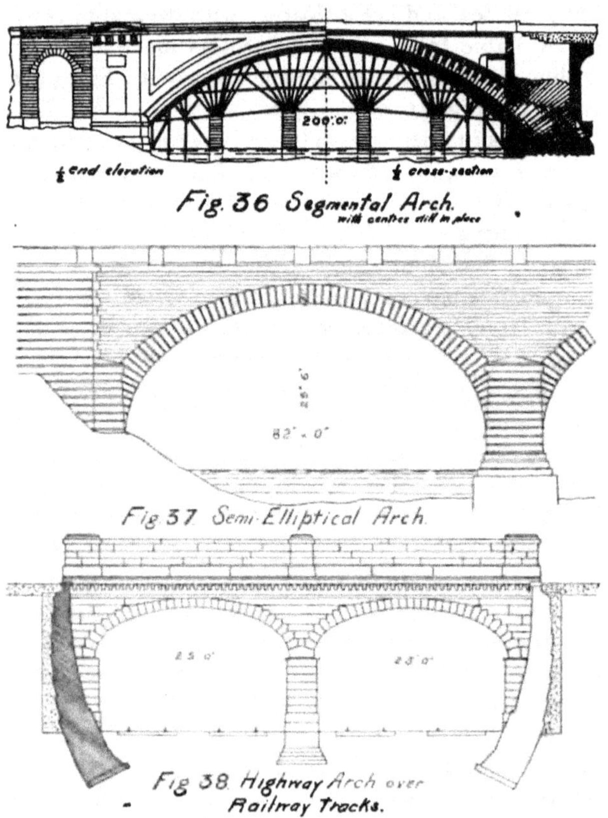

Fig. 36 Segmental Arch. with centres still in place

Fig. 37 Semi-Elliptical Arch.

Fig. 38 Highway Arch over Railway Tracks.

small rise to span, elliptical, segmental, or basket-handled, at a slight increase in cost. In small arches, it is cheaper to use roughly cut or even rubble arch sheeting of a greater depth, than to build one of first-class cut stone of less depth; but as the span increases, the economy of carefully cut and bedded arch sheeting will point to the use of the minimum depth. The workmanship on stone arch sheeting should be of the quality figured on, and if cut stone is called for, it should be as shown in the upper diagram of Fig. 40, because if left narrow at the back, the mortar that fills up the discrepancy being weaker and

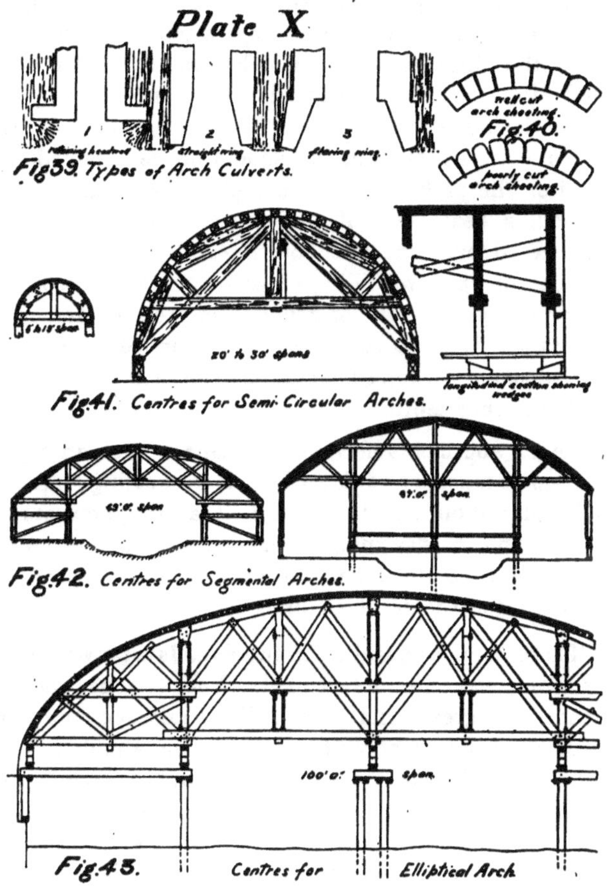

more compressible than the stone tends to throw excessive loads on the inner faces of the stones, this is point over which too great an amount of inspection can hardly be given, especially if the stones are of minimum depth. When deep rubble arch sheeting is used, the mortar will be strong enough to stand the pressures allowed.

If arch sheeting is of brick a greater depth must be allowed usually than for stone, unless the brick is of very good quality and well bonded, and as bricks are not usually made bevelled unless for a very large contract, and are all of a uniform size, the bonding of the several rings of brick in arch sheeting of several bricks in depth is not possible except at arbitrary intervals, depending on the curvature, when the outer ring is one brick thickness behind the ring inside it, at which point a header is inserted; for a circular arch this is about once every 33° of arc and is independent of the span. Longitudinally both brick and stone arch sheeting should be well bonded also, and after the arch has been completed and the centres removed, a heavy coat of cement mortar (1 to 1) should be plastered over the back of the arch down over the haunches or spandrel filling so as to prevent percolation through the joints. In construction of the arch and spandrel masonry the two sides should be carried up at about equal rates, as a heavier load on one side will tend to push over the timber centres.

(*c*) Arch abutments need not be made of such an expensive class of masonry as that of the arch sheeting. A rock-faced ashlar about equal to second-class bridge masonry is suitable, and in designing their dimensions due regard must be had to the character of the filling behind the abutments and the depth of filling over the crown.

There cannot be said to be any fixed law by which the dimensions can be determined. The various theories advanced disagree in vital points. Some take account of the horizontal thrusts tending to increase the stability of the abutments, and some do not; some attempt to allow for rolling loads, and others use only a uniform quiescent load. It may be said in general, that for small arches where there is a large margin of safety allowed, and under heavy banks where rolling loads have little effect, the dimensions as given by Trautwine are satisfactory, but where it is deemed necessary to construct a curve of pressures, Scheffler's theory of least crown thrust and neglecting horizontal forces errs on the side of safety and is easy of application; and in constructing curves of pressure taking the joint of rupture as the critical point it will be noticed

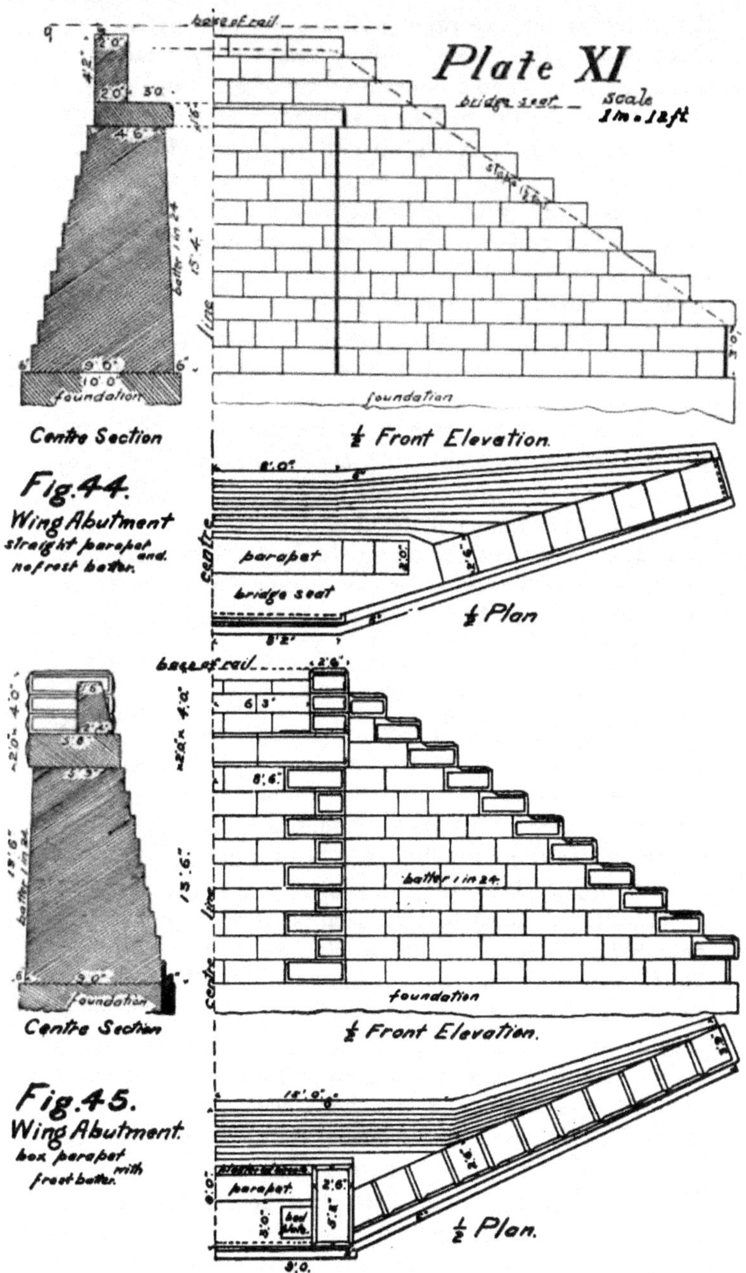
Plate XI
Scale
1 in. = 12 ft
base of rail
bridge seat
Centre Section
½ Front Elevation.
foundation
Fig. 44.
Wing Abutment
straight parapet and no frost batter.
parapet
bridge seat
½ Plan
Centre Section
½ Front Elevation.
batter 1 in 24.
Fig. 45.
Wing Abutment.
box parapet with frost batter.
parapet
½ Plan.

that these curves pass rapidly toward the back of the arch sheeting into the haunch filling before they reach the springing line, when a uniform depth of sheeting is used equal to the depth necessary at the keystone, or slightly greater; this makes it evident that there is no such thing as a semi-circular arch, but that a segment of about 120° (60° on each side of the vertical) is a true arch and the remainder is really a part of the abutment, and for this reason the spandrel (haunch filling) masonry should be carefully constructed near the springing line and of as good a class of masonry as the face walls; higher up, however, it may be of rubble masonry as its weight is its only function.

The foundations of arch abutments are a very important consideration, as a very slight settlement will derange the curve of the arch and endanger its stability. If the foundation bed is not found to be uniformly good after careful testing with an iron bar, the best way to distribute the pressure is to lay a foundation course of concrete, or piles and concrete, if the bearing power of the soil is not found sufficient

ARCH CENTRES.

To support the arch sheeting during construction and give the correct form to the arch a series of timber segments with an exterior covering of longitudinal strips or timbers is used, until the keystones are inserted. They should be designed for the weights they are to carry and rigidly supported so that absolutely no settlement may occur. But at the same time each point of support should be either on wedges that can be slackened, or on cylinders filled with sand that can be drawn off gradually at the proper time. Soon after the keystones have been inserted the easing process should begin and should be in two stages. First, a slight easement sufficient to bring each joint under pressure, and in a week or two later when the mortar has become hard they may be gradually and uniformly lowered so as to throw all the load on the arch. Figures 41, 42, 43 and 36 illustrate types of centres, and the wedges are shown in Fig. 41.

Specification for Stone Arch Culvert Masonry.—It will be laid in cement mortar of approved quality, the arch sheeting stones shall be of the length, depth and thickness shown on the plans, or such as the engineer shall prescribe, and shall be cut accurately on their intrados, beds and joints according to template, and shall

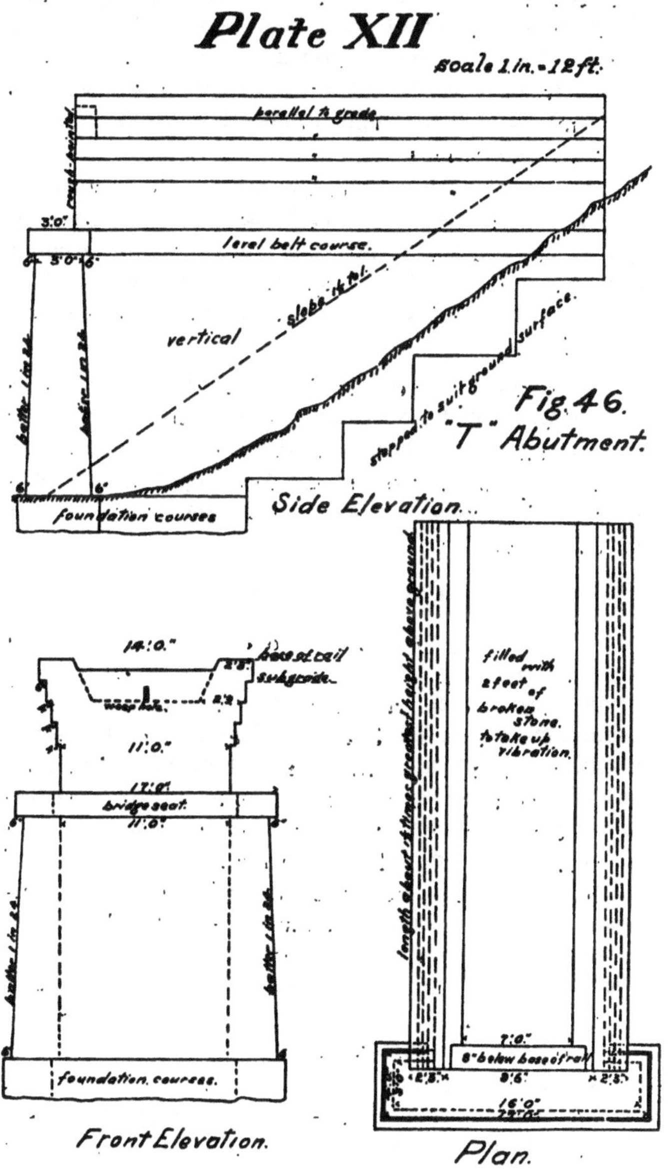

Fig. 46. "T" Abutment.

Side Elevation.

Front Elevation.

Plan.

be carefully laid with a thoroughly good bond lengthwise of the arch. The face ring stones shall be left rough on the face, except a 1½ inch intrados draft, and no projection allowed of more than 3 inches from such draft. The spandrel filling shall be rough rubble similar to good box culvert masonry, but of good bed, bond and quality of stone. The abutments, wings or head walls, and parapet walls shall be either first-class or second-class bridge masonry as the engineer may direct. (See bridge masonry specifications.

The cost of arch culvert masonry will vary with the price of stone cutting, price of brick and labor, but may be taken ordinarily at $6 per cubic yard for rubble arches; $8 to $10 for second-class arch abutments; $10 to $14 for cut arch sheeting; $8 per cubic yard for ordinary brick arches.

The quantity of masonry in arches of even the same span and rise will be so entirely dependent on the height of abutments and depth of foundations that no table will be given. The tables given for purposes of approximate estimates in Trautwine "Engineer's Pocket Book," will be found very useful.

ARTICLE 20.—ARCHES.

So much has been written on this subject and the design of large arches for rivers or roadways, or for carrying roads over railways, etc., has received such elaborate study, that it is outside the range of this work, but types of such structures are given on Plate XI., and many of the remarks on arch culverts will apply to the larger structures with even greater emphasis, (*e.g.*), the importance of immovable foundations, and taking care of the line of pressures below the haunch and in the abutments, for this purpose the abutments are often built with the beds inclined to the horizontal and nearly at right angles to the pressure, see Fig. 36, and in any case should never be further from such a right angle than the angle of friction of stone on mortar.

Whether in a given case an arch or two abutments and a plate girder span will be preferable, depends on the depth of the bank and width of stream, as far as economy is concerned, but other considerations are the greater durability and safety of an arch, its finer appearance and an absence of repairs.

The use of concrete with steel rods or wire embedded in the tension side of the arch sheeting has lately come into use for arches of small rise, especially where rolling loads tend to distort the arch, the possibility of this form of construction lies in the fact that steel and concrete have almost identical co-efficients of expansion.

ARTICLE 21.—BRIDGE SUBSTRUCTURES.

As the size of waterway increases, the cost of an arch soon becomes excessive, owing to the heavy abutments necessary for arches of long span and small height. On the other hand the cost of bridge abutments increases very rapidly with the depth of bank, so that we have two limiting features to guide us in the selection of the style of structure most suitable for a given small stream or creek, *e.g.*, with a 30-foot span and embankment 30 feet high the costs about equal each other. But whenever the arch does not cost appreciably more than the open span it should be selected, owing to the absence of floor repairs and to increased safety given. It must be remembered that the addition of a solid buckleplate floor and ballast to a plate girder will, however, make it practically safe and almost eliminate repairs.

When the stream to be crossed is of considerable magnitude the question of span lengths will be the first one to decide upon, which must be done with due regard to the probable life of iron work and the cost of replacing and painting it, as well as to the total present minimum cost of structure.

The approximate minimum cost of structure is obtained when the cost of the trusses, not including the floor system, is equal to the cost of the masonry, which should include the cost of foundations, etc., but exclude the cost of those portions of abutments of which the function is to retain the earthwork and not to support the bridge, *i.e.*, the wings, etc.; but it is usually safe to arrange the spans so that the masonry will cost slightly less than the iron, because the estimate of the latter can be made quite accurately, whereas the actual expenditure incident to river masonry construction will usually be in excess of preliminary estimates unless made with great judgment, especially in deep rivers of uncertain bottoms.

ARTICLE 22.—BRIDGE ABUTMENTS.

The most suitable design for an abutment will depend on the ground configuration, the position of the face of the abutment relatively to running water liable to scour, and to the amount of earth filling available—the various types in use are shown in Fig. 47. Of these, the tower abutment is always the cheapest, but can be used only when the embankment may be made all around it, on dry ground; the filling at the sides and front should be carefully made so as not to endanger its stability, practically it is used in two cases: First—Often as an abutment to an iron viaduct, with an adjacent heavy cutting and no stream to interfere with the foot of the slopes; and second, when a ravine is partly filled by a necessarily heavy cutting; but no borrow is convenient to complete the remainder, and an iron span must be used to bridge an insignificant stream.

The "T" abutment consists practically of a pier to carry the bridge span, and a masonry approach to carry the track onto the bridge. The objections to the "T" abutment are, that the masonry is damaged by vibration from the trains, and that the front cannot well be protected from river scour. The first objection is met by designing the stem of the "T" to be filled with two feet of rock ballast, to take up vibration; but the abutment must be kept well back from the bank of a stream liable to scour. The cost of a "T" abutment is less than any other (except the tower), when the conditions are favorable. These conditions are, that it is to carry a deck truss, and that the stem of the abutment is stepped into an ascending hillside, thereby lessening the quantity of masonry in it considerably. See Plate XII, Fig. 46. Essential advantages of this abutment are that it is practically in stable equilibrium from earth thrusts as the earth slopes run around the stem, and that water cannot lodge behind it. For through trusses needing wide piers, a "T" abutment is not especially economical, unless the saving on the hillside steps is considerable. The masonry for the pier should be first or second-class bridge masonry, but the interior of the stem may be of heavy rubble, with a cut facing, thus reducing the cost of the abutment to an average price of $8 to $10 per cubic yard.

The "U" abutment is similar to a "T," except that the stem is split into two parts and separated until considerable filling can be placed between the two parts. For deck spans up to 25 feet in height and through spans up to about 30 feet in height, the masonry in this abutment is less than in a "T" abutment, but above this height the quantities increase very rapidly owing to the increased lateral dimensions of the wings, which are designed as level retaining walls. The class of masonry necessary, however, is superior in the wings to that of the masonry in the stem of the "T" abutment, and the average cost of masonry will range from $1 to $2 per cubic yard more, or, say, $9 to $12 per cubic yard, so that it is very seldom less expensive to build than the "T" abutment, but it is very much used, owing to an impression that the wings are not affected by the train vibration. It must be kept well back from a scouring stream, and the toe of the slope in both of these classes of abutments should be protected by rip-rap, if there is any running water. Another serious objection to a "U" abutment is that it is liable to lodge water between the wings. This should always be provided for by weep holes. Whether this abutment is cheaper than a wing abutment will depend on the allowable slope of the earth, and also on the economy that can be effected by stepping the wings into the hillside, see Fig. 48. Some engineers economize masonry in the stems of "T" and wings of "U" abutments by introducing semi-circular arches of 10 ft. to 20 ft. span, just back of the pier-portion of the abutments

The wing abutment is usually used where the ground is level behind the abutment, and where the face is close to running water liable to scour, in which case the wings are flared back about 30° so as to prevent any contraction of the waterway. This abutment presents a neat appearance, and the backing may be made of rubble masonry, thus reducing the cost of the whole to about the price of "T" abutment masonry, *i.e.*, $8 to $10 per cubic yard, but it has several objectionable features: if the foundation below footings is deep, to good bottom, the quantity of masonry in the foundation is excessive. See table XV. (C). And

Plate XIII.

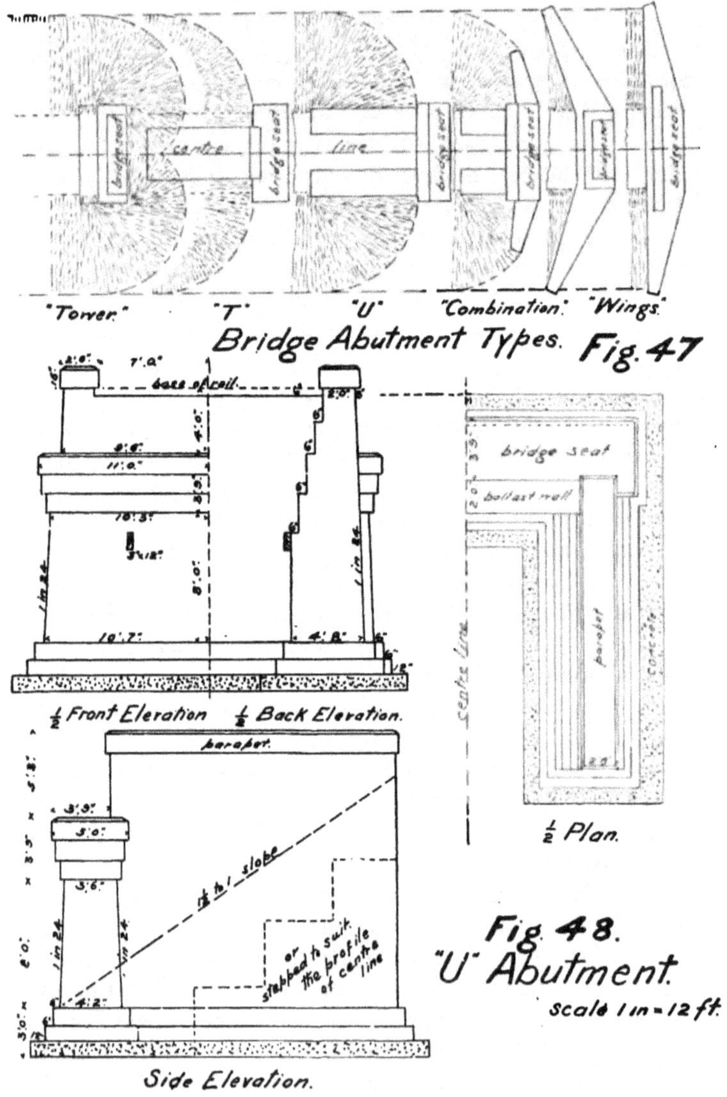

Bridge Abutment Types. Fig. 47

Fig. 48. "U" Abutment.

also, its design as a level retaining wall is always a question of more or less doubt. The ordinary rule of the width at base, being $\frac{4}{10}$ of the height + the front batter, is satisfactory if the filling behind is of average quality, but if made of heavy wet slippery clay, the structure may be in danger. Again, in designing the foundation it is necessary to know that it will always receive support in front or else the rule of $\frac{4}{10}$ must be carried down to the foundation bed. For these reasons, an abutment with a straight back and only a tapering to the wings is preferable to one with wings flared back, as it does not hold water behind it so readily and is not under so severe a strain tending to crack the wings from the body of the abutment. To increase the stability of a wing abutment the foundation pit in front should be always rammed solid with clay, or preferably concrete, up to the ground level, and in cold climates a frost batter given to the back of the parapet wall. See Fig. 45. This prevents the frost from dislodging it.

There is a great diversity of opinion regarding the correct cross-section of retaining walls in general, and in applying any rule or formula the utmost caution is necessary, because each design is a problem in itself; items to be considered are: the material to be used for backing, the manner of placing it, the slope of the natural ground behind and in front of the abutment, the drainage of the area behind the abutment, the kind of masonry and mortar to be used, etc. So many complexities would thus be given to any theoretical formula that in such designs it is, in most cases, best to be guided by past successes and failures in structures that have been similarly situated. The actual design of an abutment is very simple, the depth from base of rail to bridge seat, and the length of truss will be given, and also the distance apart of the trusses, centre to centre. Thus, deck trusses will need a bridge seat about 16 feet long and 3½ to 4½ feet wide, while through spans require bridge seats 22 to 25 feet long and 4 to 5 feet deep. The approximate quantities of masonry for different styles of abutments as given in table XV. and plotted in diagrams, will be understood as extending to a uniform foundation level, the quantity saved in "T" and "U" abutments by stepping into the hillside must be deducted, and the quantity in foundation courses added, both of which will be so much less favorable for wing abutments in any comparison involving deep foundations or steep hillsides.

TABLE XV.

APPROXIMATE QUANTITIES OF MASONRY IN ABUTMENTS.

(A)

FOR DECK PLATE GIRDER SPANS, BRIDGE SEATS 16 FEET WIDE, CALCULATED FROM DESIGNS GIVEN IN FIGS. 44, 46, 48.

Not including footing courses.

Style of abutment.	Depth from base of rail to top of footings						
	10′ c. yds.	15′ c. yds.	20′ c. yds.	25′ c. yds.	30′ c. yds.	35′ c. yds.	40′ c. yds.
Straight wing abutment	38	80	151	257	402	596	841
"T" abutment, earth slopes, 1½ to 1	45	120	227	366	538	742	979
"T" " " " 1 to 1	27	77	150	244	361	500	661
"U" " " " 1½ to 1	43	101	210	358	582	858	1,195
"U" " " " 1 to 1	31	71	144	243	391	575	800
Tower " thickness 50 per cent. of height	27	49	79	119	166	222	285

(B.)

FOR THROUGH SPANS, BRIDGE SEATS 22 FEET WIDE, CALCULATED FROM BAKER'S "MASONRY CONSTRUCTION."

Not including footing courses.

Style of abutment.	Depth from base of rail to top of footings.						
	10′ c. yds.	15′ c. yds.	20′ c. yds.	25 c. yds.	30′ c. yds.	35′ c. yds.	40′ c yds.
Wing abutment	44	104	186	294	426	588	781
"T" abutment, earth slopes, 1½ to 1	78	169	309	481	696	...	...
"T" " " 1 to 1	56	120	219	346	494	676	879
"U" " " 1½ to 1	47	107	206	386	650	...	...
"U" " 1 to 1	36	80	149	271	452	712	1,025

(C.)

AREA OF FIRST FOOTING COURSE FOR FOUNDATIONS OF ABUTMENTS AS ABOVE IN SECTION (A), ALLOWING 6″ PROJECTION ALL AROUND.

Style of abutment.	Depth from base of rail to top of footings.						
	10′ sq. ft.	15′ sq ft.	20′ sq ft.	25′ sq. ft.	30′ sq. ft.	35′ sq. ft.	40′ sq. ft.
Straight Wing Abutment	265	384	569	797	1,047	1,330	1,673
"T" abutment slopes 1½ to 1	202	299	400	499	599	700	801
"T" " " 1 to 1	142	209	280	349	419	489	559
"U" " " 1½ to 1	193	311	477	670	917	1,076	1,250
"U" " " 1 to 1	129	213	323	458	614	724	806
Tower "	114	149	196	243	290	337	384

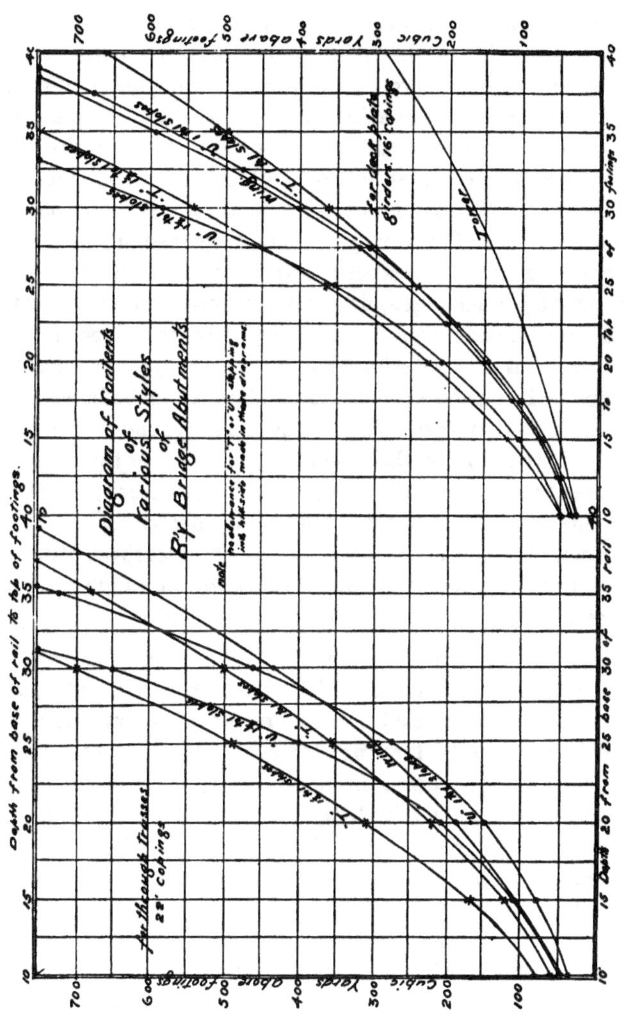
Diagram of Contents of Various Styles of R'y Bridge Abutments.
Depth from base of rail to top of footings.
Cubic Yards above footings
For deck plate Girders. 16' copings
For through trusses 22' copings
Tower

ARTICLE 23.—MASONRY BRIDGE PIERS.

The material most commonly used for masonry pier construction is stone, although in certain cases brick may be cheaper and also satisfactory; but wherever there are strong currents of water or jams of ice or logs, brick is hardly suitable, as the individual pieces are liable to displacement; but concrete piers are becoming quite common in certain parts of America, the compact and simple form of a pier lends itself readily to concrete construction, the ebb and flow of tides will not affect well made concrete whereas it has a disastrous effect often even on the heaviest class of masonry, percolating through the many joints loosening the stones and mortar; concrete piers are usually much cheaper than first-class masonry work, and even the compromise of a pier with the sides and top of stone and the interior of concrete, offers considerable saving. In designing piers located in river beds we have three special questions to study, namely, the dimensions and construction of the coping, the batter of the sides to insure stability, and the design of the cutwaters; of course the question of foundations enters into designing this class of work even more seriously than with structures built on land, but the subject of foundations will be taken up in a more general way applicable to all structures.

(1) The coping dimensions will be determined by the width apart of the trusses, the size of the bed-plates, and the loads to be carried locally on the masonry work, the variation would be for single track roads, from say 4 feet by 16 feet under coping for small deck trusses to as large as 12 feet by 40 feet for long through spans. The coping should project two or three feet beyond the bed-plates at the ends, and six inches to two feet at the sides, depending on the weight of the span. Copings should consist of an 18-inch thickness of very strong carefully made concrete, surfaced with a layer of 1 to 1 mortar before the concrete has set, or if of stones, they should be large, well-bedded, cut all over the top surfaces by fine pointing, pean hammering, or in some way giving a good surface for the bed-plates; the vertical joints should be cut and preferably the beds also; but the faces look better with the quarry face left on, if the rest of the pier is rock faced

ashlar. They should be dowelled into place, or clamped to one another, and so arranged that the pedestal plates of the bridge trusses will come *exactly* on the centre of a large stone, or if one stone cannot be found large enough to distribute the load, a large, deep pedestal block should be cut for the purpose and placed on top of the coping; for very large trusses, a steel pedestal is constructed to distribute the load over several coping stones (*e.g.*, new Victoria Bridge). Coping plans showing the exact size and position of each stone should be furnished the contractors, and it is a mooted question, whether better results as to exact surface, etc., can be obtained by bedding in mortar, or by shimming up the whole coping to an exact level on wooden chips, then pointing up all the outer bed and vertical joints, and pouring liquid grout into the receptacle of interior joints and beds thus formed until every crevice is filled, the latter plan is probably preferable, if care is taken to have the joints and beds open enough to secure their being thoroughly filled, particularly the beds.

(2) The batter of the sides is usually 1 in 12 or 1 in 24 and is a matter of appearance, as vertical piers would look top heavy, but in each case a calculation should be made for stability, under the most unfavorable circumstances; considering the stability in direction of the railway line, the forces acting would be (*a*) the wind blowing at 45° to the direction of truss, on the truss, train and pier, at say 40 lbs. per square foot, the force of a fully braked train covering one or two spans depending on the location of the expansion rollers, at say, 10 per cent. of the weight of the train, and vertical loads which would consist of a *loaded* span and the weight of the pier itself; the resultant of these forces should not fall appreciably outside the middle third of the base, (*b*) with the wind as before, but blowing on the truss and pier only, and the vertical loads of an *unloaded* span and the weight of the pier itself, the same criterion for stability should be applied as in (*a*), and in either case the remedy for instability would be to increase the batter, or introduce steps or offsets of, say, 6 inches every 10 or 12 feet, which would have the same effect. The stability of a pier at right angles to the bridge is practically never in question unless of very great height,

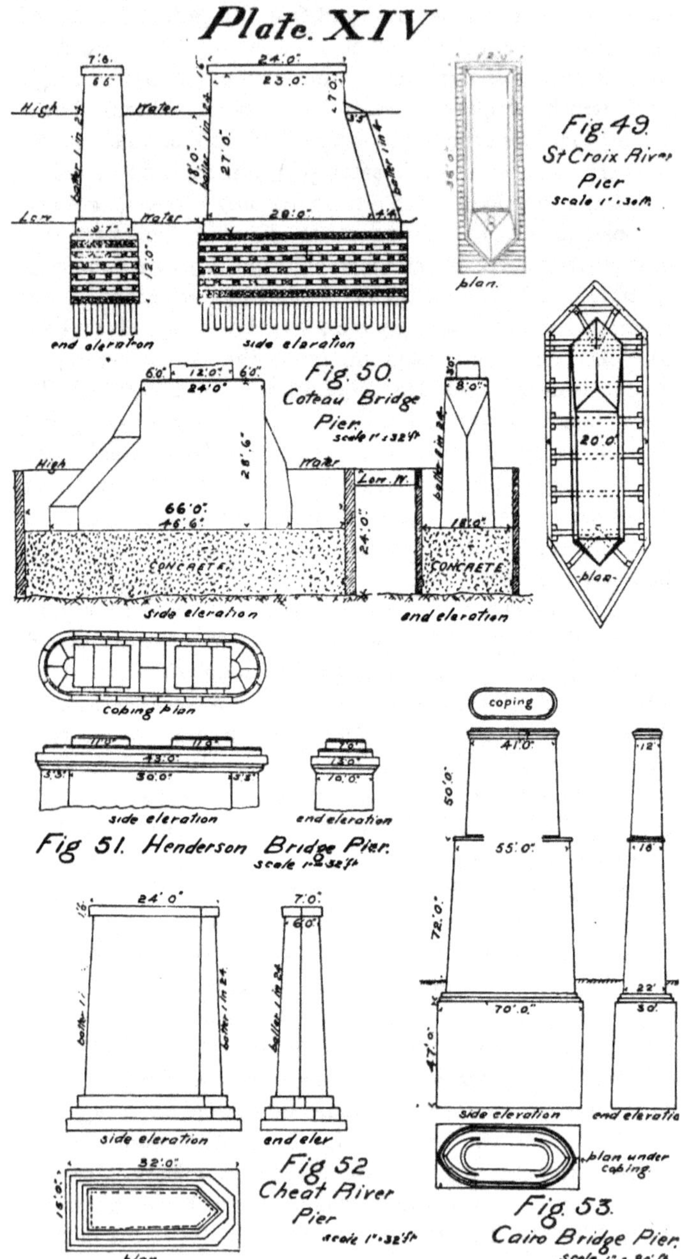
Plate. XIV
Fig. 49.
St Croix River Pier
scale 1" = 30ft.
plan.
end elevation
side elevation
High Water
Low Water
Fig. 50.
Coteau Bridge Pier.
scale 1" = 32ft.
CONCRETE
side elevation
end elevation
plan
coping plan
side elevation
end elevation
Fig 51. Henderson Bridge Pier.
scale 1" = 32ft.
side elevation
end elev
plan.
Fig 52
Cheat River Pier
scale 1" = 32ft.
coping
side elevation
end elevation
plan under coping.
Fig. 53.
Cairo Bridge Pier.
scale 1" = 80 ft.

in which case similar tests would determine what the necessary length of base should be.

(3) Cutwater designs:—Wherever there is any appreciable current in a river, it is necessary to construct the up-stream end of the pier of such a form that it will divide masses of driftwood, ice, logs, etc., as well as the current itself. Probably the simplest form is that shown in Fig. 52, which wlll not cost appreciably more to construct than a square pier, as the nose is a right angle and the faces of ordinary quarry-faced ashlar, but such a form is suitable only for streams carrying light ice or moderate jams of logs; in place of this the more ornamental forms shown in Figs. 51 and 53 would be equally satisfactory, especially the latter, but cost considerably more, and should therefore be used only on very important structures.

Where piers are to be placed in swift currents, or in any stream carrying heavy jams of logs, or thick floes of ice, their cutwaters should be of designs similar to Figs. 49 or 50. The cutwater of the former hardly extends high enough and is not flat enough for Canadian rivers, and besides it lacks the valuable addition of a small pointed lower end, which is introduced to eliminate an eddy at that point in swift currents, which tends to undermine the end of the pier, unless on solid rock. Probably the St. Croix pier and cutwater are suitable for the conditions they were designed for, but the intention of the design on Fig. 50 is that the jams will rise on the nose and split in two, passing on harmlessly.

Stone masonry bridge piers will cost from $9 to $15 per cubic yard, depending on their height, size and the cost of quarrying, transporting and cutting suitable stone. If expensive cutwaters are needed this will add to the cost; those used under severe conditions being of a first-class cut stone construction of large dimensions, clamped together and dowelled also, with a strip of boiler plate added to the nose to prevent dislodgement of stones. The following specification of first-class bridge masonry will apply to both abutment and pier construction, but in many cases a less severe specification has resulted in satisfactory and durable work.

Specification for First-Class Bridge Masonry.—"This class of masonry will be ranged rockwork of the best description, from stone of approved weathering qualities, and will be laid in suitable cement mortar (1 to 2 natural, or 1 to 3 Portland). The face stones will be accurately squared, jointed and bedded, and laid in courses not less than 12 inches thick, decreasing in thickness from bottom to top of walls; the joints and beds to be less than half-inch and joints well broken, no break to be less than nine inches. The stretchers to average at least 3½ feet in length with 3 feet as a minimum, to have at least 16 inches bed, and always at least as much bed as rise. The headers to have a width of not less than 18 inches, and to hold the size back into the heart of the wall that they show on the face; they shall occupy at least one-fifth of the area of the face of the wall, and be practically evenly distributed over it, so that the headers in each course shall divide equally or nearly so, the spaces between the headers in the next course below. When the walls are not more than 3½ feet thick, the headers shall run entirely through, and when between 3½ and 6 feet thick, there shall be as many headers of the same size in the rear as the front of the wall, and the front and rear headers must alternate and interlock at least 12 inches with each other. In walls over 6 feet thick, the headers shall be at least 3½ feet long, alternating front and back as just described, their binding effect being carried through the wall by intermediate headers of a similar character. The stretchers in the rear of the wall, and the stones in the heart of the wall shall be of the same general dimensions and proportions as the face stones with equally good bed and bond, but with less attention to vertical joints, and must be well fitted to their places, and carry the course evenly quite through the wall; a header shall in no case have a joint directly above or below it, but rest entirely on a stretcher at the face; any small interstices that may remain in the heart of the wall shall be carefully filled with mortar and spauls. The face stones shall be left rough on the face, with no projection of more than three inches from pitch lines, and two-inch drafts will, in general, be carried up and around all projecting angles.

In the construction of piers, it is understood that the description above given for face work shall apply to both ends and both sides of the pier. Copings are to be cut and dowelled or clamped according to coping plans furnished, the top shall be crandalled and pean-hammered, or otherwise brought to a smooth surface, and so arranged as to bring the pedestal plates of the trusses exactly on the centre of especially large coping stones of dimensions given on the plans."

ARTICLE 24.—METAL OR METAL AND CONCRETE PIERS.

On Plate XV. are shown several applications of meta and concrete for bridge piers. The method by screw piles founded in mud is quite unique, and has not been attempted in America for railway bridge piers, but the steel cylinders filled with concrete and founded either in mud on piles or anchored to the rock by iron dowels are more familiar. For light highway bridges this method is quite suitable, but it is probable that, except in the form of one large cylinder, as a cofferdam to a pneumatic caisson, both filled with concrete and sunk in waters not needing cutwaters, their use for railway bridge piers will be exceptional, as the more massive forms shown on Plate XIV. will be better able to withstand the vibrations of trains and impacts of ice, etc.

ARTICLE 25.—IRON VIADUCTS OR TRESTLES.

The features usually under the control of the railway engineer are the general lay-out, the design and construction of masonry, and the erection of iron work; the detail designing of the iron being essentially a branch of bridge-work.

(*a*) *General lay-out.*—This has resolved itself in America to be, in general, a system of braced independent towers and suspended spans (see Fig. 54); the towers are usually about 30 feet spans, with posts vertical in side elevation; and, in end elevation, the girders are spaced 8 to 10 feet centres, and the posts battered at 2 to 3 inches per foot, depending on the allowance for wind, the aim being to avoid tension in the windward pedestals at the most unfavorable instant. The suspended spans are usually also plate girders of 30 feet to 60 feet span, depending on the height, the greatest economy being claimed when the cost of girders and longitudinal bracing is equal to the cost of towers and pedestal masonry. The usual design

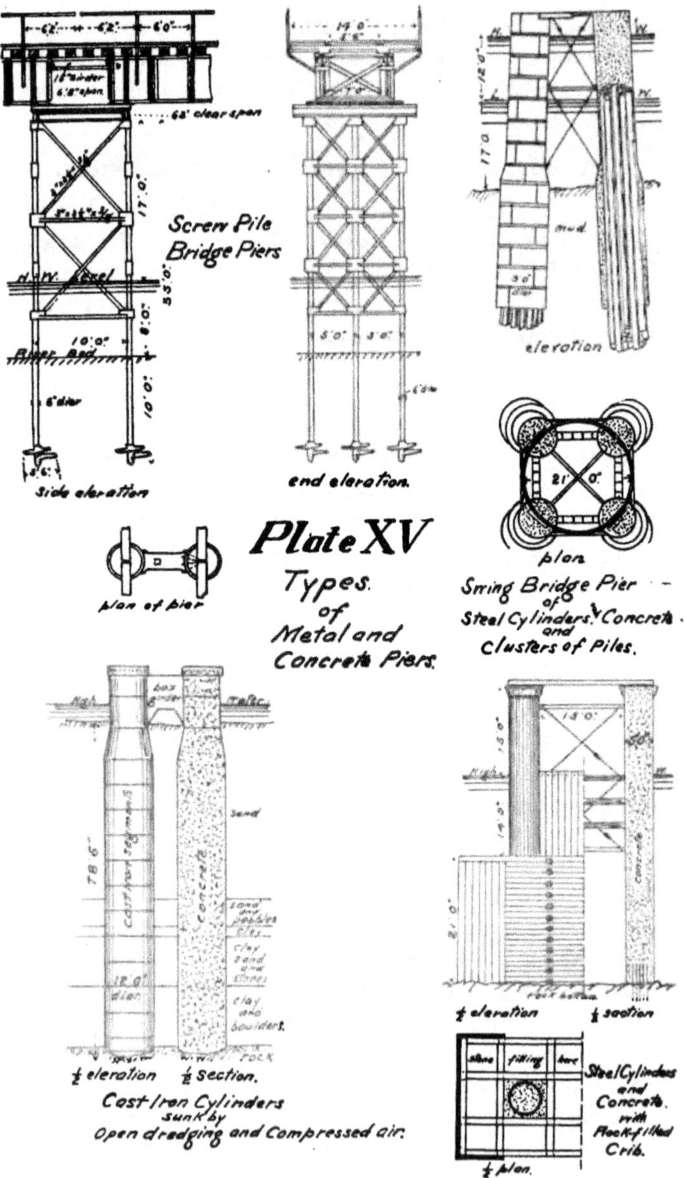

is with diagonal rods acting in tension and the girders resting on top of the posts, with slotted holes for expansion, but some late designs are for rigid, riveted bracing and posts extending to the tops of the girders, which are riveted to the webs of the posts, temperature changes being taken up in expansion pockets every 100 to 200 feet. This system is theoretically more rigid, but costs more and demands a perfect system of pedestals, any settlement being dangerous to a proper distribution of stress.

In estimating the weight of iron for approximate work, the following rule may be useful:

The weight of metal in the towers and bracings, in pounds, is equal to about 8¼ times the longitudinal section area in square feet of the ravine below the line of girders and between the faces of abutments; this is based on a 100-ton consolidation engine and a high viaduct, say 100 feet, this will be changed to say 9½ times for a viaduct 50 feet high, and 10½ for a viaduct 35 feet high with the same weight of engine. The weight of girders, in pounds, may be estimated at $(9l+100)$ pounds per foot run of a span, where l = length of span. The price of iron varies considerably, with cost of erecting falsework, if any, and cost of erection and freight in general, but is about four cents per pound, in place, as a minimum. The floors of viaducts usually consist of say 8 inch by 10-inch oak ties about 12 feet long on their edge, boxed one-inch over the girders, and fastened to them by hooked bolts which pass through the guard rails and hook under the upper girder flanges; the spacing should be not more than six inches clear, and guard rails either double or with an inner guard rail of ordinary flanged rail. Some recent designs, however, call for solid floors of steel troughs filled with ballast and with ordinary track ties, which, certainly, would lessen vibration and increase safety in case of slight derailments.

(*b*) *Pedestal Masonry*.—The greatest care must be exercised in laying out and building the abutments and pedestals, as any error in position or appreciable one in height is very serious, because the iron work, being manufactured at a distance contemporaneously with the masonry, is shipped to the spot partially assembled, and cannot be afterwards altered except by a slight shimming up of pedestals built too low. The shoes of the columns are always bolted down to the copings, and in high structures these bolts should be built into the pedestals 5 or 6 feet, by passing through the stones as they are laid in place (see Fig. 54), the bolt holes are afterwards filled tight with sulphur, lead, or neat cement grout, while if the structure is not very high or subject to heavy winds, the anchor bolts are only sunk into the coping stones, dovetailed by wedges and cemented as before. Pedestal masonry is sometimes built of well burnt brick covered by a stone coping, and the use of monolithic concrete is especially adapted to this class of work, as it permits of the anchor bolts being buried in the concrete to any depth during construction. In any case the best class of work must be done, as the strains and thrusts acting are of higher intensity than in ordinary masonry, and the pedestal is not only under considerable vibration for so small a mass, but may be put in tension on the windward side of high structures. The cost of stone pedestal masonry will vary with the conveniences for handling small quantities of large stones at each spot, and with the amount of cutting required, from $10 to $15 per cubic yard, whereas a very high class of concrete may be done for at the most $8 per yard as there is nothing to move from place to place but the plank moulds and mortar box. The coping of concrete pedestals should be made very strong, say, 1 cement, 1½ sand, 3 stone, with a layer of 1 to 1 mortar for 1½ or 2 inches thick on the top and sides, put on at the same time as the rest is being built, and incorporated with it by vigorous trowelling and ramming, etc.

ARTICLE 26.—WOODEN TRESTLES.

In America where timber is plentiful, cheap and widely distributed, where capital is often limited and construction hurried, and in inaccessible regions, the use of wooden trestles has been of great economic value. It is yet a factor not to despised; and although their design is apparently not very difficult, yet it should be thoroughly understood by the young railway engineer in its many phases at an early date in his practice. We have probably 100 feet per mile of railway in Canada, or about 300 miles altogether, and we may safely figure on a continual renewal of at least half of this every seven or eight years, which represents an annual drain on our forests for this purpose of perhaps 40,000,000 F.B M., while the other half

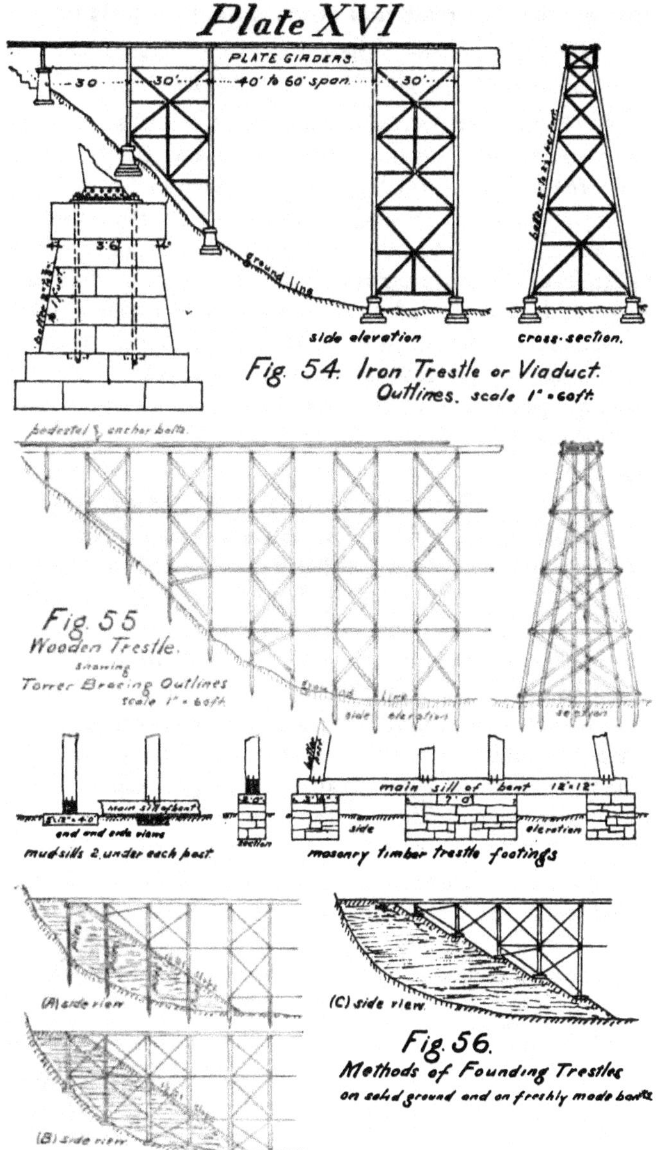

Fig. 54. Iron Trestle or Viaduct. Outlines. scale 1" = 60 ft.

Fig. 55 Wooden Trestle. showing Tower Bracing Outlines scale 1" = 60 ft.

mudsills 2 under each post

masonry timber trestle footings

Fig. 56. Methods of Founding Trestles on solid ground and on freshly made banks

will be replaced by earth, stone or iron as fast as it decays. Gradual separation and sizes of timber available have produced several distinct types, each of which has its advocates; the greatest diversity exists in every detail from floor to foundation, and will be discussed somewhat, in order to obtain a broad view of the various methods in use.

(*a*) *Foundations.* — These may be classed as (1) masonry footings, (2) piles, and (3) mudsills.

(1) Stone work suitable for culvert walls is used for the masonry footings and will cost about $4 per cubic yard; the heaviest and longest stones should be used for the top course, and the sill well bedded on it in mortar. It is not necessary to bolt the sills to the masonry if the timber bents are braced longitudinally; and, sometimes, where it is anticipated that an iron viaduct will replace the wooden one at the first renewal, it may pay to build the batter post pedestals under every alternate bent, of first-class pedestal masonry, which will serve as pedestals for the columns of the iron viaduct when needed (see Fig. 56).

(2) When foundations are deep or soft, or less expense is desirable, a pile is driven under each post of each bent, and the sills rest on and are drift-bolted or tenoned to the piles; for 15-foot bents and a 100-ton engine, a pile will carry one-quarter of the load on a bent when it penetrates less than about two inches under a blow from a 2,000-pound hammer dropping 25 feet; but if such a penetration cannot be obtained without great depth of pile-driving it may pay to use more piles. (See Fig. 55.) Whenever a trestle is less than 15 feet high, framed bents may be dispensed with, and the piles left driven up to the bottom of the caps on which the stringers rest, in which case the outer piles are driven to a slight batter, and then pulled in, as the cap is drifted on, to a total batter of two or three inches per foot, or if the bent is only five or six feet high the piles may be all driven vertical; in either case, the centre piles are five feet apart and directly under each rail; usually a pair of × or sway-braces is bolted on to each bent, but no longitudinal braces are necessary It is not good practice to build these pile trestles over 15 feet high unless for temporary work, as the vibration

from traffic loosens the piles, and the rot at the ground surface is rather rapid and wasteful of material cut off, in case of renewals.

(3) *Mud Sills or Bank Sills.*—These consist of six or eight short cross-sills under the main sill of each bent, the usual dimensions are 8x12 inches by 5 feet long laid on flat, unboxed, which facilitates shimming by means of planks or boards under the main sills, and the bents will not get out of plumb if longitudinal braces are used. (See Fig. 56.) Other special foundations are:—Cribwork filled with stone where there is a current of water, a grillage or floor of timber in case of soft submerged ground, and holes cut in the rock, carrying the ends of the posts without any sill on steep solid rock side hill.

(*b*) *The founding of end bents.*—This is generally a difficult problem, as the freshly made embankments at each end of the trestle settle rapidly. Fig. 56 shows the three methods in ordinary use. Of these (*A*) is undoubtedly the best, but most expensive, as pile driving on sloping banks costs considerable; such piling should be done after allowing the bank to settle as much as possible, and he piles should be well driven into the solid ground to prevent them being shoved out of place longitudinally. (*B*) This method is pernicious, the embankment always contorts the trestle, no matter how well braced, the posts rot off at the ground surface, and it is also expensive and often delays the construction of the embankment; it has little to recommend it, unless the whole trestle is to be soon filled in. (*C*) This is seemingly a flimsy method, but is found very satisfactory in mild climates at least. Each bent is supported on eight or ten mudsills, and as the bank settles a bridge gang surfaces and relines the trestle by shimming between the bents and bank-sills; after a few months, when the bank has become firm, very little attention will be required, and it is always convenient for repair or renewal, and nothing rots but the mudsills. It has also the especial merit of providing an elastic and gradual approach from the rigid trestle to the sinking fresh bank, which the first and second methods do not present. It is the cheapest method, and its only demerit is that it is affected by frost in the spring in cold climates.

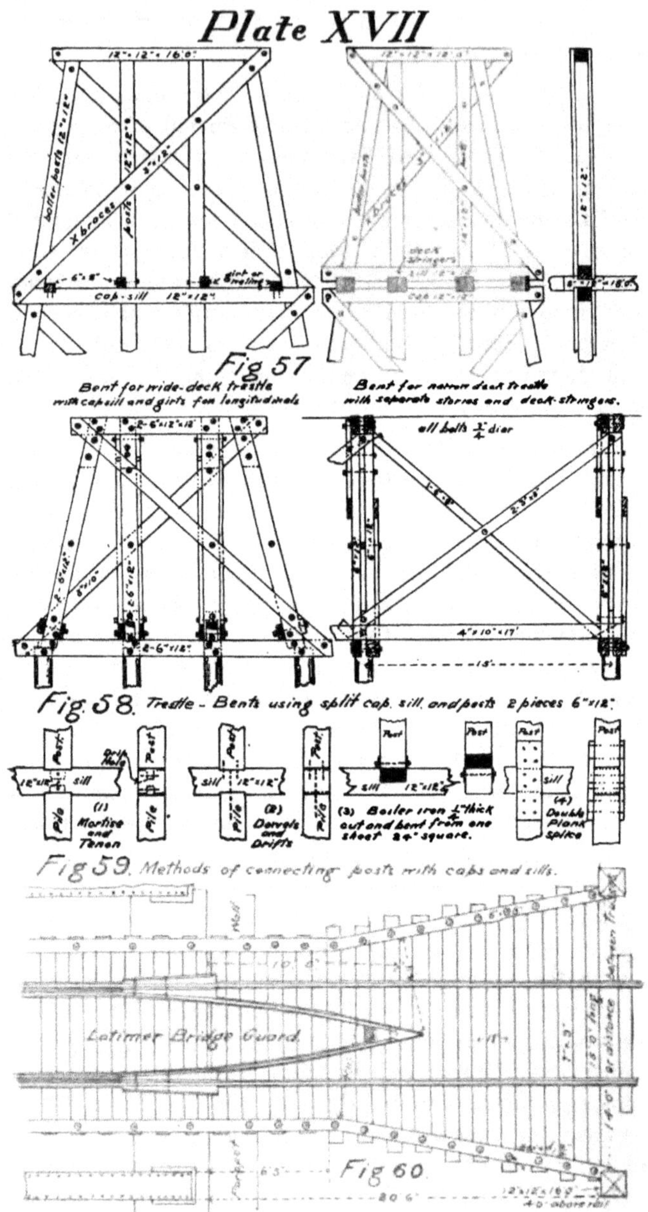

Fig 57

Bent for wide-deck trestle with capsill and girts for longitudinals

Bent for narrow deck trestle with separate stories and deck-stringers.

Fig. 58. Trestle-Bents using split cap, sill, and posts 2 pieces 6" × 12".

Fig 59. Methods of connecting posts with caps and sills.

Fig 60

(*c*) *Trestle Bents and Systems.*—The ordinary styles of bents are with four posts, which are increased to six posts in very high trestles, each story or deck being braced by one or two pairs of cross or sway braces—the plumb posts being five feet apart, and batter posts sloping at from two to three inches per foot.

The longitudinal braces consist of diagonal braces and horizontal ties. The former are so grouped as to form towers and spans alternately, but the horizontal ties run the entire length of the structure. Speaking generally, low trestles on tangents need very little longitudinal bracing, while high ones on sharp curves need very much; between these two, the extent, disposition and sizes of such bracing will be a matter for the judgment of the designer, but all such braces should be bolted and not spiked, as vibration from trains will soon loosen the latter, and even bolts from the same cause and from shrinkage of timber need to have their nuts tightened two or three times before the rust locks them into place. Sway braces may, however, be partially bolted and partially spiked with pressed spikes.

The posts, caps and sills are usually of solid timbers 10 inches by 10 inches, 10 inches by 12 inches or 12 inches by 12 inches in section, the latter being most common. (See Fig. 57.) And such timbers make the best structures; but large timbers are often hard to obtain at reasonable prices, and in such cases, the style shown in Fig. 58 is found satisfactory and is easy to repair and renew. The split caps, sills and posts are usually 6 inches by 12 inches well bolted together; but the cost of such bolting, and danger of some of the bolts working loose are objections to the system. On the other hand, water is not so apt to lodge or rot where timbers are so narrow. An extension of this system is the "cluster bent" trestle, the posts being of four pieces of 6 inches by 6 inches, breaking joint and bolted to split caps and sill, with the cross braces acting also as separators. This style is advisable only in localities where timber is obtainable only in small dimensions, and of poor quality, its advantage being facility of repair; while the trestle itself is not so stable or durable as if made with larger timbers.

(*d*) *Joints and Fastenings.*—Fig. 59 shows the best methods in use. The advantage of mortise and tenon work is the easiness with which repairs can be effected, but it is a favorite lodging place for water and one of the first points to rot in a trestle—a drip hole partly obviates this; mortise and tenon work costs about $1 per M.B.M., more to frame than other methods shown. The drift bolt and dowel method is superior to the first one in rigidity, durability and easiness of erection, but is hard work to tear down, while the third method, although not in common use, appears to be a very sensible joint; the fourth method is for temporary work only, and is fastened with 6-inch pressed spike which would work loose very soon; it is an expensive and clumsy joint, but saves timber from mutilation for a second using. Probably the best combination is a tenon at the top of posts and 2 dowels 8 inches by 1 inch diameter at the bottom, such a system avoids rot and enables posts to be easily renewed and replaced.

(*e*) *High Trestles.*—When trestles are higher than 20 to 25 feet more than one story or deck will be needed (see Fig. 57), these may be either in separate bents locked by deck stringers, or they may be in a continuous bent in which the sill of one bent becomes the cap of the one below, etc.; in the former case the four lines of deck stringers, about 8 inches by 12 inches, overlap and are boxed onto the caps and sills and gained into them, giving a lock joint. The structure is very rigid, is simple of erection and easy to repair, but needs more material than the latter method, in which the longitudinal bracing consists of four lines of girts or walings about 6 inches by 8 inches, which are butt-jointed and boxed about 3 inches on to the caps or posts and also bolted; this method saves one cap and some timber in the longitudinals also, and the upper and lower sway braces may be fastened by the same bolt at the cap-sill, but the trestle is a little harder to erect and much harder to repair, it is, however, on the whole probably the preferable method of the two.

(*f*) *Floor Systems of Trestles.*—The floor consists of stringers, ties and guard rails, and should be completed by adding some form of bridge guard like that shown on Fig.

60, by which trucks not too far off centre may be brought back to direction and probably placed again on the rails, or, at any rate, carried safely across the structure.

In Canada the wide-decked trestle, with two main nests of stringers and two jack stringers supporting ties 12 feet to 14 feet long, is generally used on standard gauge roads, partly to give room for snowplows inside the guard rails and partly for additional safety; this is nowadays further supplemented by two lines of guard rails, of which the inner ones are faced with angle iron fastened by countersunk screws.

The sizes of guard rails and ties are shown in Figs. 61 to 65; but it is a growing feeling that ties cannot be spaced too close for safety to prevent bunching in case of derailment. They should not be more than 6 ins. apart clear, at most, and preferably 4 ins., and every fourth or fifth tie should be spiked to the stringers, and all boxed from ½ in. to 1 in. on to them. There are some, however, who claim that if ties are kept spaced by the guard rail, and are boxed down on to the stringers, that any further fastening is unnecessary; ties should be of oak or some durable hardwood, which will hold track spikes.

The guard rails should be of pine, or some durable wood that will not warp, and boxed down about 1½ inches on to the ties, so as to hold them apart, as well as to act as guard rails. They should be bolted on outer ones and spiked on inner ones to about every fifth tie. There are various styles of stringers; that shown on Fig. 61 is now almost standard for Canada. It consists of alternate groups of two and three stringers in two nests, one under each rail. These stringers are long enough to lap one foot and rest twelve inches on the caps, and are bolted together, but separated by cast washers to prevent rot, and each nest is drift bolted down to the caps. The outer or jack stringers carry very little of the load and are butt jointed, and also drift-bolted to the caps. The style shown in Fig. 62 is practically obsolete, being considered rather top heavy and demanding too heavy timbers; the corbels also give a very uncertain aid to the strength of the stringers. The style shown in Fig. 63 is used considerably in the United States. The stringers are 32 feet long, alternating with butt joints, and make a very stiff floor; but such large timbers are hard to obtain and harder to take in and out during renewals. One advantage, however, is that all ties will have the same boxings, while in Fig. 61, the length of boxing

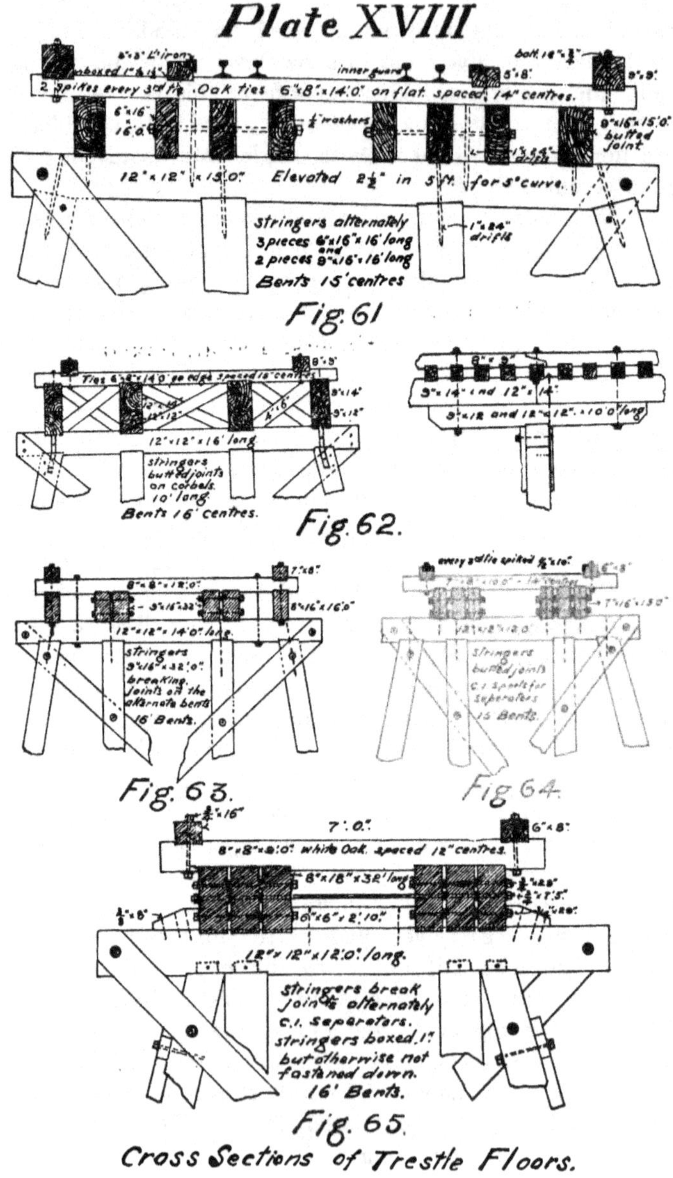

Cross Sections of Trestle Floors.

varies with each alternate bent—but the method of fastening the ties to the caps by long bolts is objectionable, such bolts will spring loose, and the heads would be cut off by the wheels of a derailed train. In the greater part of the United States, the narrow decked trestle, such as shown in Figs. 64 and 65 is very much used, there is an evident economy, the bents being narrower, the jack stringers omitted and only one guard rail, and that one close to the steel rails. It is claimed that such trestles with rerailing guards at each end, close spaced ties, and with an inside steel rail guard, are as safe as wider ones, and that a train which would derail so badly as to get outside the inner guards would leave the trestle anyway. Such floors are sometimes drift bolted down, or as in Fig. 65, held in place by a system of braces and rods—such a method having the advantage of facility of renewals.

(*g*) *Elevating Trestle Floors on Curves.* — This will amount to from a half-inch to five-eighths inch in five feet per degree of curve, and may be accomplished in various ways which are given in order of merit.

(1) Framing the bent with cap inclined to the required slope.

(2) Inserting an elevation or secondary cap on top of a level bent cap, and notching it down to suit the curve.

(3) Notching the main cap on light curves to receive the stringers.

(4) Placing blocks longitudinally under the ties and on top of the stringers.

(5) Sawing specially sloping ties.

(6) Placing strips or blocks under the steel rail but on top of the ties.

The sloping ties are, apparently desirable, but in renewals such a varied assortment is needed for trestles on various degrees of curve that it is an intolerable nuisance.

(*h*) The cost of trestle work is about as follows:

Cast-iron washers, etc...2c. per lb.
Iron bolts, etc. ....5c. per lb.
Framing and erection ..$6 per M.B.M. (drift bolted).
" " ..$7 " " (mortise and tenon).
Cost of timber ....$17 to $27 per M.B.M.

A very ordinary price for timber now being $30 per M.B.M. in place, including iron; but in designing structures of this nature, by having due regard to the dimensions of timber which can be most readily obtained in any locality, the price may be held down, often somewhat lower than this, while sometimes a special price is put on stringers, which should be designed for the loads intended, and carefully inspected to see that they are up to the requirements, as they are the only members of the structure ever likely to be strained to their limit.

(*k*) *Quantities of Timber in Trestles.*—For very important and high structures, tables are of little value; but for ordinary use the following will give quite approximate quantities:—

TABLE XVI.

APPROXIMATE QUANTITY OF TIMBER IN TRESTLES PER 100 LINEAL FEET FOR VARIOUS HEIGHTS.

Narrow Deck, Timbers 10 in. x 10 in.			Lake Erie and Western.		Wide Deck, Timbers 12 in. x 12 in.	
Height.	F.B.M.		Height.	F.B.M.	Height.	F.B.M.
15 feet	17,000	1 story.	17 feet	19,400	10 feet	25,700
25 "	22,000	1 story.	22 "	22,000	20 "	30,500
35 "	25,000	1 story.	27 "	26,650	30 "	38,600
45 "	34 000,	2 stories.	32 "	29,640	40 "	47,000
55 "	44,000	3 stories.	37 "	34,780	50 "	55,200
65 "	50,000	3 stories.	42 "	38,080		
75 "	56,000	3 stories.	48 "	43,920		

ARTICLE 27.—FOUNDATIONS.

The foundation of a structure is much more important than any other feature of its design, as on its security depends that of the structure itself. We may therefore be justified in examining closely into the bearing power of soils, the various methods of increasing it or distributing the load over greater areas, and the various methods of sinking or making foundations.

The bearing power of soils varies very much, being least for soft clays and unconfined quicksands, and increasing through sand, gravel, firm clay and hardpan to solid rock; it may be as low as one-half ton per square foot, and as high as 180 tons per square foot, but it is only with the lower values that we need deal. A good clay foundation will bear safely 2½ to 4 tons per square foot, and dry sand or gravel from 5 to 6 tons, while cases have been known of 10 to 11 tons per square foot being carried. In railway work 3 or 4 tons per square foot is usually all that we want to put on to a foundation; consequently, whenever good firm clay (which will not be watersoaked), sand or gravel are reached, the foundation may be considered safe, provided we have sounded several feet, and in case of important structures, many feet below the foundation bed, to see that the substrata are equally firm. Rankine's earthwork formula takes account of the depth of the foundation beneath the ground level, and one rule in use in the Western States is that the safe load per square foot on sand is 2 tons + 1 ton for each 10 feet in depth. No account has been taken of the friction of the soil on the sides of the structure below ground level. This is only of importance where the foundations are sunk by pneumatic caisson or open dredging, in which case it is variously estimated at from 200 to 600 pounds per square foot of surface exposed to side friction, if the caisson is of cast iron, about 300 lbs. for wrought iron, 900 lbs. for masonry, 300 to 600 lbs. for timber, and for piles 500 to 1,000 lbs. per square foot.

FOUNDATIONS ON LAND.

Where a firm foundation is obtainable within a few feet of the surface, all that is necessary is to dig a pit below frost level, or to any required depth and commence stonework at once, or often lay 1 or 2 feet of concrete to get a uniform bed to lay on, if the foundation is bouldery or uneven; but if, on the other hand, the foundation bed shows soft spots or uneven bearing power, the difficulty may be overcome by using a deep bed of concrete which will span over small spots (provided there is no danger of undermining by scour from an adjacent river, liable to change its channel), or by using a grillage of timber instead, in cases where the timber will always remain watersoaked, but not otherwise. Again, the same expedients may answer if the whole foundation is somewhat, but not very, soft, by spreading out over a larger area. When, however, it is found that the foundation is too soft for such methods even if thoroughly drained, and that it will not pay to dig down far enough to reach a firm one, recourse must be had to some artificial method of obtaining greater bearing power, the usual way being to drive piles, although in some cases where cost is no object or the piles would be subject to decay, large platforms formed of cross-layers of steel I beams or rails bedded in concrete (as in Chicago), have been used to great advantage, the transverse strength being such as to enable a very much enlarged area being used. When sufficient piles are driven to carry the load required they may be cut off level and capped by drift-bolting on a double layer of 12-inch by 12-inch timbers, as a base for stonework, or preferably either on the broomed heads of the piles or after sawing them off, a layer of concrete is placed and also rammed around and between the piles, in which case the earth between the piles will take a portion of the load, being much compacted by the driving, indeed, in cases where timber is scarce, a method has been adopted of pulling out piles after driving and immediately filling the holes with well rammed sand, the compacted earth and the sand pillars together carrying the load safely. Cast iron and wrought iron screw piles are used extensively in ocean jetty, lighthouse, and pier construction, but for foundations to masonry structures on land the wooden pile is in

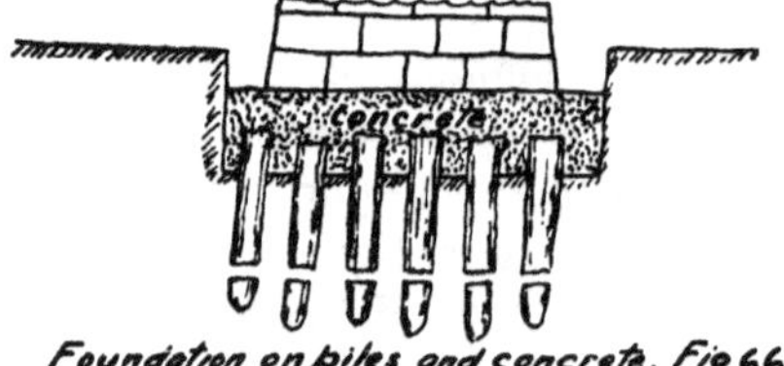

Foundation on piles and concrete. Fig 66

almost exclusive use; of the various ways of driving piles by steam hammer, drop hammer, water jet, gunpowder explosions or insistent weight, the first three are the only ones worth considering.

The water jet is used economically where water is plentiful, ordinary pile driving inconvenient, and in sandy, quicksandy or silty soils (indeed it is often the only means of driving piles in bad quicksands), the water is carried in a pipe down the side of the pile and is projected in large quantities and considerable pressure just below the point of the pile; this allows the pile, assisted by a dead weight or by light blows from a small hammer, to sink very rapidly, the water rising in a film around the surface of the pile and almost eliminating friction as long as the action continues, but after driving is finished the sand or earth settles around the pile and gives as high a bearing power as with hammer-driven piles.*

The Steam Hammer weighs from 4,000 to 5,000 lbs. and sits on the head of the pile, striking a blow with a piston loaded with about 3,000 lbs., the stroke is 2½ to 3 feet in length and about once per second. This keeps the soil and pile in a continual vibration and effects more than the occasional, though more severe blow of a drop hammer. Although a steam hammer is unable to drive in very hard ground economically, and is not economical where only a few piles are to be driven in a place, as the cost of transportation of hammer, boiler, steam-pipes, etc. is too great, yet wherever a great many piles are to be driven in one locality, as on docks, large foundations, etc., or where the outfit can be economically transported from place to place, it can drive piles much more cheaply and quickly than a drop hammer.

The Drop Hammer is raised either by hand, horse, or steam power, and usually a trip is arranged to free the hammer automatically at the top, the line being brought down again by hand, which wastes time, in order to avoid which, sometimes, in driving by steam, the line is permanently attached, and a friction drum is utilized to let it drop without tripping, thereby dragging the line with it and lessening the force of the blow, but economizing time. The chief danger in this method lies in its abuse by dishonest contractors, where the pile-driving is specified for certain maximum penetration at the last blow, in which case, a slight friction on the drum will materially lessen the effect of the blow. The weight of drop hammers varies from 1,200 to 2,000 lbs., and the drop from 15 to 30 feet depending on the hardness of driving and pile head, and length of leads.

The drop and steam hammer are in general use, and the load which a pile can safely carry may be approximately estimated by various empyrical formulæ, when the drop, weight of hammer, and penetration at the last blow (with an unbroomed head to the pile) are known, such formulæ are usually on the safe side,

*See *Engineering News*, Vol. 31, 1894, page 316, for jet pile driver.

because although at least one instance has been given of a pile refusing to drive at all, and yet after a day's rest going five or six inches at a blow, yet the usual experience is, that piles which will penetrate several inches at a blow, after long continued driving, will penetrate less, or almost refuse to drive, after a rest of a day or two allowing the material to settle around the pile.

The three formulæ looked upon most favorably in America are:—

(*a*) Weisbach's or Sanders—

$$\text{Safe load} = L = f \times \frac{12\,WH}{S} \quad (f = \tfrac{1}{3} \text{ to } \tfrac{1}{8}).$$

(*b*) Trautwine's—

$$\text{Safe load} = L = f \times \frac{46\,W\sqrt[3]{H}}{S+1} \quad (f = \tfrac{1}{3} \text{ to } \tfrac{1}{12}).$$

(*c*) Wellington's—

$$\text{Safe load} = L = \frac{2\,WH}{S+1} \quad \text{(drop hammer).}$$

$$\text{Safe load} = L = \frac{2\,WH}{S+\frac{1}{10}} \quad \text{(steam hammer).}$$

(L = safe load in lbs.; W = weight of hammer in lbs.; H = drop of hammer in feet; f = factor of safety; S = penetration, per blow, in inches, for average of last four or five blows.)

Where penetrations are as much as two inches or over there is not much to choose between all these formulæ; but for small penetrations the first one is evidently inapplicable, giving abnormally high results; on the other hand, the third formula is conservative under all conditions, simple in use, and admits of a modification for steam hammer driving, in which *stiction** is not an element to be considered. It is also applicable down to zero penetrations for ordinary weight of hammer, drop and length of pile, as $L = 2\,WH$ is about what a pillar 12 or 15 inches in diameter and, say, 20 feet long will safely stand (taking W = 2,000, H = 25).

These formulæ neglect many small losses of energy in driving, and are, therefore, only empyrical; but the results accord fairly well with the few facts known regard-

---

* Stiction is that excess of frictional resistance offered, when a body is started from rest, over the continuous frictional resistance offered to a body while in motion, under the same conditions of surrounding material. C. B. S.

ing the safe loads on piles. Evidently in ordinary soils, the skin friction is the important element in their sustaining power, as the load carried on the point of even a blunt ended pile would not be very great unless on solid rock.

Of the ordinary losses of energy in pile-driving, the only important one (neglecting the friction of the leads, the compressing of the pile, the bouncing of the hammer, which latter can be remedied by lessening the drop, or getting a heavier hammer), is the brooming of the head and point under hard driving. To prevent the former, a heavy iron band about three inches by one inch should be fitted around the head, and when brooming does occur, it should be sawed off at once, while in hard or bouldery ground the point should be shod with straps of iron and made rather blunt—in fact some drive with piles almost without a point, claiming that brooming is thus prevented—in quicksand, a pile turned butt downward will often be the only means of keeping it down during driving, and in all cases where piles are subject to severe vibration they will take a much less load than where it is a quiescent one. This is not included in the formulæ and must be provided for in the design.

For permanent piling in fresh water districts such woods as cedar, oak, yellow pine, rock elm, spruce and tamarac are in common use, being given somewhat in order of merit for durability, while any good wood may be used for temporary work. The cost for ordinary lengths will vary from five cents to eight cents per foot in favorable localities, to twelve cents or fifteen cents for oak piles with a longer haul, and even twenty-five cents or thirty cents per foot for the same in great lengths of fifty or sixty feet; driving will vary from two cents or three cents per foot in situations where great quantities are driven by steam hammer, to four cents or five cents per foot where driven by a track pile driver on cars; but for ordinary railway construction work, scattered and in small quantities, eight cents to twelve cents per foot is not out of the way. An ordinary price on railway construction is about thirty cents per foot for good piles, 12 inches in diameter at the small end, in place in the work, and fifteen cents per foot for all cut-off ends, this includes all labor of driving and cutting off to the exact height if on land, but cutting off under water is extra, and sometimes a matter of considerable expense.

Deep Foundations on Land. — Where foundations require to be carried down for any considerable depth, it is more economical to timber and sink vertically than to put earth slopes on the foundation pit, such timber will,

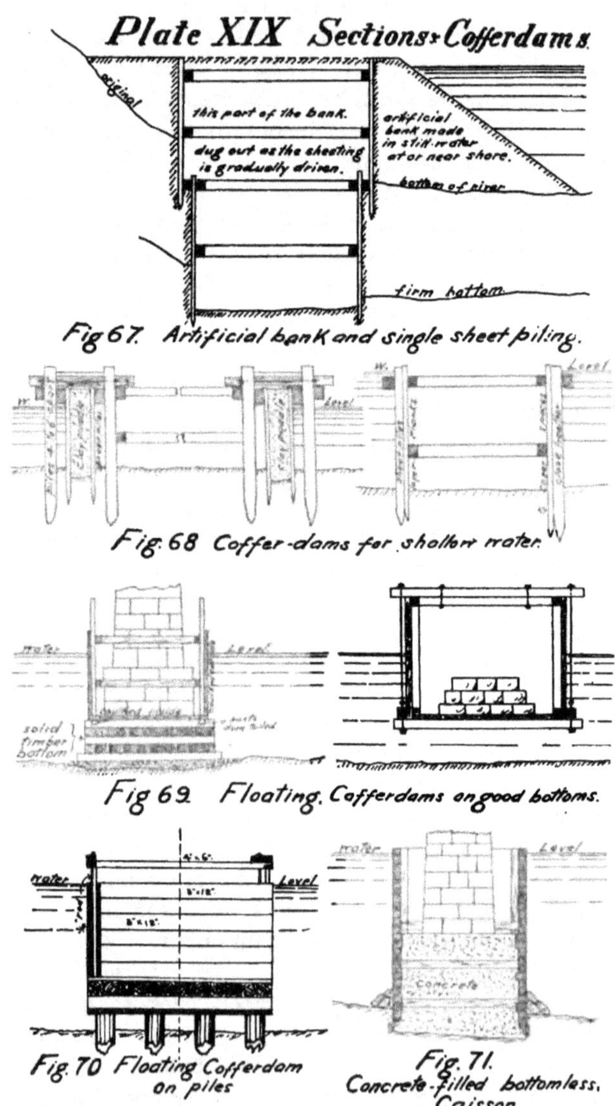

Fig 67. Artificial bank and single sheet piling.

Fig. 68 Coffer-dams for shallow water.

Fig 69. Floating. Cofferdams on good bottoms.

Fig. 70 Floating Cofferdam on piles

Fig. 71. Concrete-filled bottomless. Caisson.

in general, consist of vertical hand-driven sheet piles and horizontal rings of timbers and braces to sustain them at five or six foot intervals, as shown Fig. 67, Plate XIX., and in planning such work it is well to remember: (1) To allow for extra room more than is apparently required, in order to give freedom of movement, extra timbers, etc.; (2) To be sure to find out how deep it is to good bottom before digging is begun, in order to see whether piling might not be less expensive, and chiefly so as to be able to say definitely how many rings of timber there will be, and how much extra room will, therefore, be needed to step in all around about 15 inches every 10 or 12 feet in depth; this is a very important consideration. See Fig 67.

FOUNDATIONS IN WATER.

When masonry work is to be built in water, the considerations which determine the method to be adopted are: (1) The depth of water and its fluctuation in level. (2) The depth of soft material underlying the water which must be penetrated to secure good foundation. (3) The velocity of the current. (4) The money and materials available.

Of these considerations Nos. 1 and 2 are most important, and the total depth from water surface to bottom of structure will determine whether the foundation should be obtained by:

(*A*) Fixed cofferdams; (*B*), floating cofferdams, with solid timber bottoms; (*C*), bottomless cofferdams or caissons; (*D*), compressed air; (*E*), open dredging.

Fixed Cofferdams.—These are used where there is shallow water, at most only a moderate current, and where the bottom is of such a nature as to admit of sheet piles being driven in and the foundation suitably excavated to a firm foundation bed. This may be accomplished in various ways. Where the structure is near shore and waste excavation is available, it will pay to make an embankment above water level, and carry down hand-driven sheet piles kept in position by rings of timber, the excavation being always kept about level with the bottom of the sheet piles. This method is illustrated in Fig. 67. If, however, the structure is not thus situated, the sheet piles are either driven in a double layer, as in Fig. 68, or a single row of Wakefield sheet piling is used (this is an artificially made sheet pile composed of three planks spiked together to form a tongue and groove). If any of these methods are employed, a centrifugal pump will be kept busy keeping down the water while the foundation courses are being laid, but in case such pumping power is not advisable or available a more expensive form of cofferdam can be used, as in Fig. 68. Rows of guide piles are driven at considerable distances apart, then walings are bolted on, and a double row of sheet piles is driven around the area to be unwatered, between which is rammed clay puddle, making a very watertight, but expensive, cofferdam. This is generally employed in extensive works where the area is to be unwatered for some length of time.

Floating Cofferdams with Solid Timber Bottoms — It is moderately certain that as long as timber is covered with running fresh water it will never decay, and it is even contended that in any fresh water it is practically safe also; this has led to the adoption of methods of foundation building which do not involve the unwatering of the bottom. If the bottom is bare and moderately level, or can be dredged to a good bottom and levelled up with broken stone, it is manifestly easy to build a watertight box with either a solid timber (Fig. 69) or stone-filled crib, or only a plank layer, as a bottom (Fig. 69), and, after floating it into position, sink it, by building in it or by external loading, and after the structure has been built up above water level tear off the sides of the watertight box, leaving the bottom as a permanent part of the structure. If, on the other hand, the foundation is soft and good bottom can be reached by piling, the piles are driven to a firm bearing, sawed off under water close to the bed of the river, and the same operation as just described is gone through, the structure being landed on top of the piles as a foundation, as in Fig. 70. These methods are cheap and satisfactory in situations where the current is not excessive, but in very swift currents such constructions are not as manageable as the bottomless cofferdams to be described, and even where used, it is found advisable to build the timber work and footing courses of masonry considerably (1 to 2 feet) larger than the neat work, which is laid out after the crib is in its final position. This provides for some permissible inaccuracy in sinking. The cribs are well drift-bolted together and the boxes caulked with oakum and dove-tailed or bolted down to the bottom, so as to prevent them lifting when the sinking process is going on.

Bottomless Caissons or Cofferdams.—Where no timber is desired under the masonry, or where the current is very swift, the method shown in Fig. 71 has been found best, but is only admissible where good foundations are easily obtainable. The bottomless box is floated into place, loaded until it sinks to the bottom, and then is either unwatered by having a large canvas flap around the outside of the bottom, held down by bags of concrete, thus nearly sealing the bottom, the caisson being then pumped out and the bottom excavated or levelled off with concrete, or else, if the bottom is already firm, as is usually the case in swift currents, there is no necessity for unwatering until a great depth of concrete has been put in, forming a watertight bottom ; in the latter case, if there is an irregular rock bottom, the caisson cannot be made to fit it, and in order to keep the undertow from carrying away the concrete as fast as deposited, or at least dissolving out the cement, it is found necessary to fasten a canvas flap around the inside of the bottom and load it down with bags of concrete, pea straw, etc., until a bottom has been formed And in depositing the concrete it is done by lowering an iron box, with a hinged bottom, containing about one cubic yard down to the bottom ; the box is tripped, allowing the concrete to slide gently out, whereas it would become dissolved if allowed to fall any distance through water. After such a bed of concrete has been formed as is considered sufficient, the caisson may be pumped out and construction continued in open air.

Compressed Air.—Where a great depth of water and soft foundations are encountered the methods previously described must be abandoned. Early in this century the vacuum air process was tried, by which the excess of outside pressure forced soft materials up inside a vacuum chamber, this material being excavated, air was again extracted, and each time the hollow chamber of wood or iron sank down by its own or added weight ; but this method was found uncertain in its means of directing the sinking, was capable of only limited application and failed entirely on encountering stiff clay or boulders, besides it did not enable the bottom to be personally examined and properly prepared for the foundation layers.

Very soon the plenum or compressed air process was tried, and to-day it is recognized as being in every way most satisfactory until greater depths than about 100 feet below water level are to be obtained, when open dredging through wells must be resorted to. Figs. 72, 73 and 74 show common forms of the same process. The drawings are almost self-explanatory, the pressure of air in the working chamber is constantly maintained, and the extent of the pressure must always be sufficient to keep out water; the tendency being for compressed air to be continually escaping around the working edges, and bubbling up to the surface outside the chamber. Where pneumatic cylinders are used, they are in pairs, sometimes braced together, the two supporting one end of a truss, and being completely filled with concrete after bottom is reached. See Plate XVI. One larger cylinder, as in the Hawkesbury bridge, with elliptical ends will, however, be much more stable.

Where large timber working chambers are used they must be very strong, as the whole weight of the pier will be carried on their backs until the working chamber is filled in, which is not until firm bottom is reached. It may be shod with iron or merely with timber, depending on the materials to be met with, and on top of this chamber may first be constructed a timber crib as in Fig. 74, extending up to the ground surface and filled with alternate pockets of concrete or broken stone sufficient to sink the chamber, which crib is built up gradually as the process goes on. Or, if advisable, the masonry may be commenced immediately on top of the working chamber as in Fig. 73, this will usually be done where the foundation is not a very deep one. The support which a deep caisson sunk by this method, or by open dredging gives to a pier and bridge, is partly by the bearing on the bottom and partly by friction on the sides, which is estimated at from 300 to 600 lbs. per square foot of surface, and is an enormous item in such a structure as that of Fig. 74, amounting to 2,000 or 3,000 tons. Of course, this resistance is not all to be overcome while sinking, for the continual movement and escape of a film of compressed air tends to aid sinking by lessening friction.

Plate XX

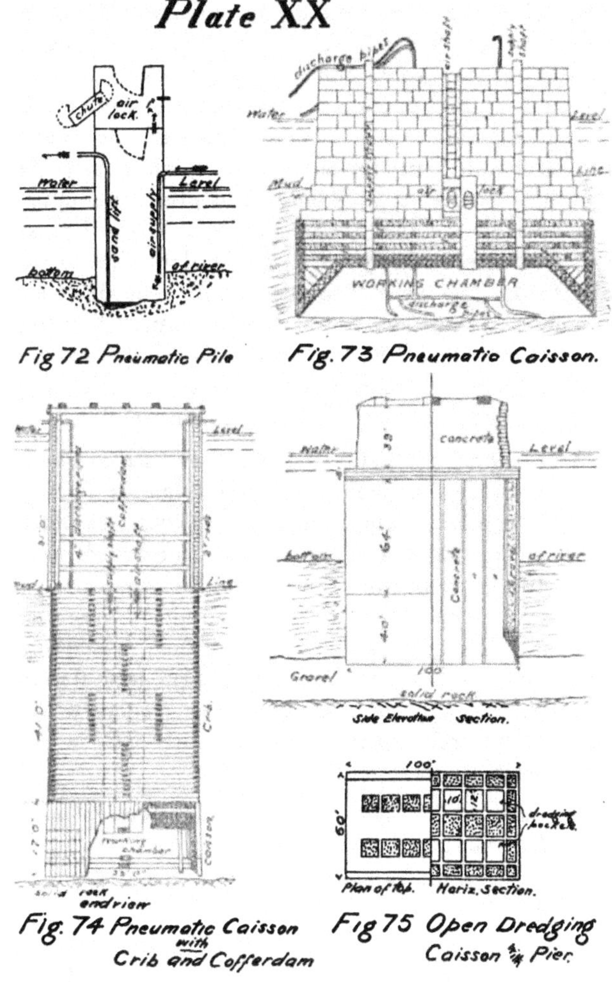

Fig 72 Pneumatic Pile

Fig. 73 Pneumatic Caisson.

Fig. 74 Pneumatic Caisson with Crib and Cofferdam

Fig 75 Open Dredging Caisson for Pier.

The material to be excavated is forced out of the discharge pipes by the compressed air, if it is finely divisible, by opening valves at the mouths of flexible pipes, but boulders, gravel, logs, etc., must be laboriously taken out of the air lock in small quantities, making the operation costly. The air shaft and lock form the means of ingress and egress, and it is a question whether it is safer and more convenient to have the air lock near the top or bottom of the shaft, the former, however, being safer for the men. The process of working the lock is to open one door, pass in, close the door, open a valve so as to raise or lower the pressure as the case may be, and then open the other door and pass on, some time being necessary to prevent injury to the lungs and ear drums; men can work in about 4 atmospheres pressure as a safe maximum, and then only for three or four hours for healthy men; with less pressure the period of labor may be lengthened, but on coming out to the open air the depressing effect of a lowered pressure must be counteracted by a strong stimulant like coffee, to prevent injurious consequences; and for reasons of safety the compressed air is taken from a receiver and not direct from the compressor. So that an accident to the machinery might not have an immediately disastrous effect by permitting an inrush of water before the men could escape.

The supply shafts marked are only used at the last, where concrete is passed in through them, to fill up the working chamber. The shafts themselves are also all filled with concrete, and the whole structure is a solid mass of timber, concrete and stone. Sinking foundations by compressed air has many advantages—the sinking can usually be quite accurately directed; it enables all the pier construction and excavation to proceed together; it enables all kinds of materials to be removed, and it permits of a careful examination and preparation of the bottom before concrete is put into place.

An example of such construction is detailed in Patton's "Foundations," at a cost of

$16.82	per cubic yard	for the	caisson	material;
$10.76	"	"	crib	"
$7.83	"	"	sinking	"

making an average of about $20 per cubic yard. This was for a depth of 68 feet below water and in 56 feet of mud—evidently the cost would vary with the depth and the materials encountered, in this case the sinking was at the rate of 1½ to 2 vertical feet per day.

Open Dredging.—This method has been long practiced in India, where circular brick wells are built with heavy walls, and gradually lowered through soft soils by excavating and undermining, the material being raised in some primitive manner. This is improved upon now by using steel cylinders and excavating by clam-shell or other dredges. There is usually difficulty in controlling the direction of movement, and large logs and boulders are troublesome, so that open dredging is usually used only where the depth is too great to admit of using compressed air and where the materials can be freely dredged (in India, submarine blasting of boulders, etc., tended to crack the cast iron cylinders, but would, probably, not have a bad effect on timber cribs). A striking example of this process is that of the foundations for the Poughkeepsie bridge, which were sunk through a depth of 50 to 60 feet of water and 75 to 80 feet of mud, clay, sand and gravel, or a total distance of 140 feet below high water. (See Fig. 75.) The cribs, 100 feet by 60 feet, had 31 gravel pockets, extending from top to bottom, which afforded enough load to sink the cribs when undermined; there were 14 dredging wells extending from top to bottom 10 feet by 12 feet, in cross section, through which the dredging was done by a clam-shell dredge. The walls were 2 feet thick of solid timbers, laid in alternate lapping courses, well drift-bolted together, and after the cribs arrived at good bottom the wells and pockets were filled with concrete, and a floating caisson similar to Fig. 69 was brought out into position and built into until it sank on to the top of the crib. The chief difficulties in carrying out this process, aside from anchoring such a huge mass of timber in a swift current, preparatory to dredging, are that it is difficult to guide the crib in direction as it settles down, and that logs, boulders, etc., under the cutting edges, cause delay and necessitate, often, sending down divers. A combination of compressed air and open dredging has been used in Europe, in which several working chambers surround an open well, the men force the material out under the inner edges of the working chambers from which it is removed by a clam-shell dredge; the process is cheaper in handling the material, but uses an enormous quantity of compressed air.

ARTICLE 28.—LAYING OUT AND MEASURING WORK.

Cross-sections should be taken at such intervals that the prismoid between two adjacent ones will have planes as boundaries or a top surface with longitudinal convolutions only, extending in straight lines from one section to the other; to do this quickly and without unnecessary sections is a matter of experience and visual judgment, requiring the personal attention of the engineer. The

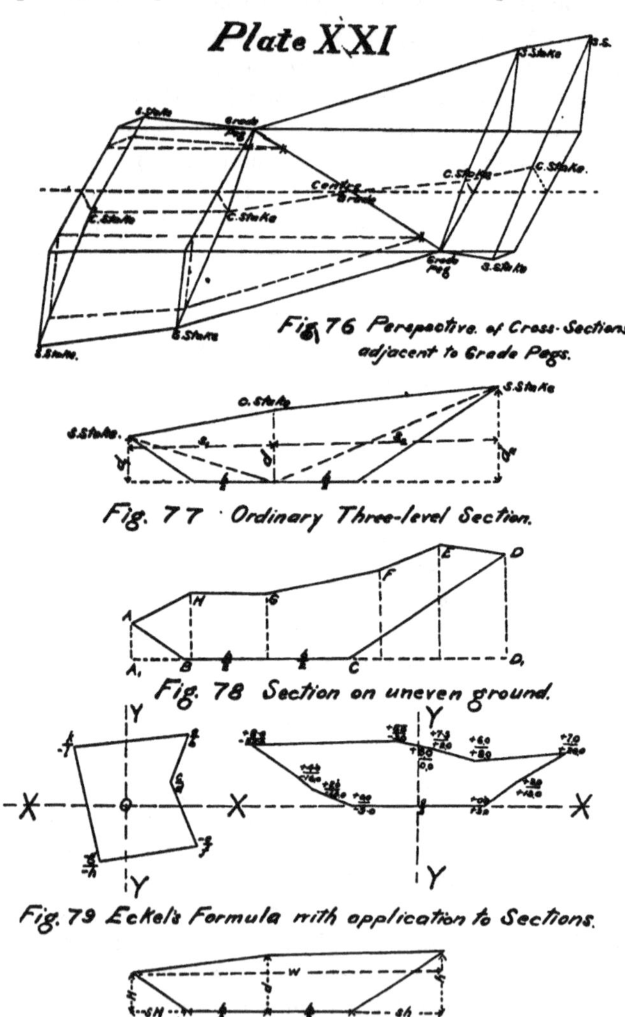

Plate XXI

Fig. 76 Perspective of Cross-Sections adjacent to Grade Pegs.

Fig. 77 Ordinary Three-level Section.

Fig. 78 Section on uneven ground.

Fig. 79 Eckel's Formula with application to Sections.

Fig. 80 Prismoidal Formula Diagram.

slope stakes should be marked on one side with the cut or fill and on the other with the distance from the centre line; some engineers also write the station (chainage) on the slope stakes. These stakes are put in at every 100 feet for light work and on tangents, but on curves and heavy work they should be put in every 25 or 50 feet, depending on circumstances, and on all side-hill work, liable to slip, the sections should be carried up the hillside 200 or 300 feet to points beyond any danger of movement, and should be taken before excavation has been commenced.

There are two methods of keeping notes in use in Canada; in the first, each rod reading is entered in a separate line and the corresponding cut or fill reduced from the grade elevation; in the second method the difference between height of instrument and grade is called "grade rod," and the rod readings are subtracted mentally from it, and the corresponding cuts or fills are recorded, consecutively, on one line of the book in the form of fractions, with the distances from centre line as denominators. It is evident that the first method is more laborious and fills much more space in a note book, and is not so convenient for plotting, but, on the other hand, the reductions can be checked afterwards, and are legal documentary evidence, whereas the second method is entirely one of convenience and leaves great chances for error by careless mental subtraction, which cannot be duplicated, and the note books are, therefore, not very strong evidence in a law court.

The following are notes of surface levels of Figure 79 taken by both methods:

(I) ENGLISH METHOD.

Station.	B. S.	H. of I.	F. S.	Int. S.	Ground.	Grade.	Cut.	Fill.	Remarks.
102 ...	..	311.20	..	10.2	301.0	293.0	8.0	..	..
2 R..	..	..	..	10.7	300.5	..	7.5	..	..
8 R..	..	..	..	12.2	299.0	..	6.0	..	..
20 R..	..	..	..	11.2	300.0	..	7.0	..	S. S.
3 L.	..	..	..	9.7	301.5	..	8.5	..	..
22 L..	..	..	..	10.2	301.0	..	8.0	..	S. S.

(II) UNITED STATES METHOD.

Station.	B. S.	H. of I.	F. S.	Grade.	Grade rod.	Left.		Centre.	Right.		
102 ..	...	311 20	...	293.0	18.2	(S.S.) $+\frac{8.0}{22.0}$	$+\frac{8.5}{3.0}$	$+\frac{8.0}{0.0}$	$+\frac{7.5}{2.0}$	$+\frac{6.0}{8.0}$	$+\frac{7.0}{20.0}$ (S.S.)

The greater simplicity and convenience of the second method commend it to general use in spite of its deficiencies.

On some roads it is customary to take sections after the work is done, and pay for the actual quantities excavated; in others, the slope lines are adhered to and every endeavor is toward having the full section taken out. It is probable, however, that the former method is more satisfactory to all concerned, although giving a little extra work to the engineering staff.

Structures need to be staked out twice, once for the foundation pit and again for the laying out of masonry, and of course, on all important structures, measurements and levels are given very frequently as the work progresses, both as an aid and a check to the contractors. A separate note book called a "structure book" should be used, and in it recorded, from day to day, notes of actual sizes, heights and measurements of all structures, and a duplicate office copy also kept up to date for fear of losing the field book. This structure book should include notes on all timber, stone and iron structural work, and will be a valuable aid in case of disputes involving quantities, it will also enable a large scaled profile being made for the maintenance department showing the exact chainage and elevation of each foundation and portion of structure. The necessity for absolute accuracy in laying out, measuring and calculating the quantities in all structures and grading cannot be too firmly impressed on the young engineer; all office calculations should be made in duplicate and preferably by different persons, as one cannot be very sure of a check on one's own work.

Cross Section Areas may be calculated in at least three or four ways, (1) where only a three-level section is taken, as in Fig. 77, the area of the section is evidently made up of triangles and is:

$$\text{Area} = \left(d \times \frac{S_1 + S_2}{2}\right) + \left(\frac{b}{2} \times \frac{d_1 + d_2}{2}\right) \ldots\ldots\ldots (1)$$

(2) On rougher ground where more than three surface readings are necessary this method fails and must be replaced by more tedious ones, quite ordinarily Fig. 78 illustrates the one adopted, and consists in taking out the

sum of all the trapezoidal areas, $A.A._1D._1D.E.F.G.H.A.$ and deducting the area of the two triangles $A.A._1B.$ and $D.D._1C.$

(3) By careful plotting, irregular areas can be taken out quite accurately with a polar planimeter.

(4) By Eckel's formula, which can be used without plotting the sections, and is equally adapted to the easiest or most difficult rectilinear areas. This formula, which is mathematically correct, is: "If the corners of any rectilinear polygon be referenced by rectangular co ordinates to any origin, then if the ordinate of every corner be multiplied by the abscissa of the next corner, and so on around the polygon, and these products added together; and if the ordinate of every corner be again taken and multiplied by the abscissa of the next corner, passing around the polygon in the *reverse* direction, and these products added together, then the area of the polygon is equal to one-half of the difference of these two sums." As an example, in Fig. 79 the area of the polygon is

$$\tfrac{1}{2}\{[(a \times d) + (c \times f) + (-ex - h) + (-g \times -l) + (k \times b)] - [(a \times -l) + (k \times -h) + (-g \times f) + (-e \times d) + (c \times b)]\} \qquad (2)$$

or,

$$\tfrac{1}{2}\{(a \times d) + (c \times f) + (e \times h) + (g \times l) + (k \times b) + (a \times l) + (k \times h) + (g \times f + (e \times d) - (c \times b)\} \qquad (3)$$

In which great care must be taken to use the correct plus and minus signs. In railway work, this is much simplified by having all the area above the axis $XX.$, and in very irregular areas, which are met with in cuts that have slipped in as in Fig. 79, the area can be quickly taken out, thus, as follows:—

$$\text{Section} = \frac{+8.0}{-22.0}.\ \frac{+8.5}{-3.0}.\ \frac{+8.0}{0.0}.\ \frac{+7.5}{+2.0}.\ \frac{+6.0}{+8.0}.\ \frac{+7.0}{+20.0}.$$

$$+\frac{3.0}{+13.0}\ \frac{+0.0}{+9.0}.\ \frac{+0.0}{0.0}.\ \frac{+0.0}{-9.0}.\ \frac{+2.0}{-14.0}.\ \frac{+4.0}{-16.0}.\ \frac{+8.0}{-22.0}$$

$$\text{Area} = \tfrac{1}{2}\{-24.0 + 16.0 + 60.0 + 120.0 + 91.0 + 27.0 - 32.0 - 88.0 + 128.0 + 56.0 + 18.0 - 60.0 - 56.0 - 12.0 + 24.0 + 187.0\} = 227.5 \text{ square feet.}$$

Thus arriving at a correct result without plotting sections, and by a mechanical sort of process, which is a safe one to place in the hands of a comparatively unintelligent rod-man; for the purpose of checking calculations, it is not appreciably more or less rapid than by taking out areas by method. (2)

QUANTITIES.

The use of tables and diagrams is a great aid in taking out approximate quantities, so that in various handbooks may be found the volume of 100 foot prismoids of level sections, of various heights, slopes and widths of road-bed, and this has been extended in Wellington's earthwork diagrams, etc., by giving the volumes of 100 foot prismoids where the sections, although not level, are of the three-level type, having a separate height at the centre and each slope stake, and as in easy sections this is all that is taken the diagrams are very useful. More accurate calculations of volumes of excavation or embankment may be made in three ways: (1) The prismoidal formula, which is the only one that is mathematically correct, is as follows:—

$$\text{Volume} = L \times \frac{A + 4A_1 + A_2}{6} \qquad \ldots\ldots\ldots\ldots\ldots\ldots (4)$$

Where L = length of prismoid A and A_2 = end areas and A_1 = middle area (which must be calculated by interpolating the middle heights).

A proof of this formula may be found in any mathematical text book, but a neat adaptation of the formula for three-level sections is given by Mr. G. H. White in *Engineering News*, April, 1895 (see Fig. 80).

$$\text{Volume} = \tfrac{1}{6} L \left\{ A + 4A_1 + A_2 \right\} =$$

$$\tfrac{1}{6} L \left\{ \left(\frac{W.d}{2}\right) + \left(\frac{b}{2} \cdot \frac{H+h}{2}\right) + \left(\frac{W_2.d_2}{2}\right) + \left(\frac{b}{2}\frac{H_2+h_2}{2}\right) + 4\left[\tfrac{1}{2}\left(\frac{W+W_2}{2} \cdot \frac{d+d_2}{2}\right) + \frac{b}{2}\left(\frac{\frac{H+H_2}{2} + \frac{h+h_2}{2}}{2}\right)\right]\right\} \quad \ldots\ldots (5)$$

$$= \tfrac{1}{6} L \left\{ \left(\frac{W.d}{2} + \frac{W.d}{2} + \frac{W.d_2}{2} \right) + \left(\frac{W_2 d_2}{2} + \frac{W_2 d}{2} + \frac{W_2 d_2}{2} \right) + \frac{b}{4}\left(H+H_2+h+h_2+2H+2H_2+2h+2h_2 \right) \right. \quad \ldots\ldots\ldots\ldots (6)$$

$$= \tfrac{1}{6} L \left\{ W\left(d + \frac{d^2}{2} \right) + W_2\left(d_2 + \frac{d}{2} \right) + \tfrac{3}{4} b(H + H_2 + h + h_2) \right\}$$

$$\ldots\ldots\ldots\ldots\ldots\ldots\ldots\ldots\ldots\ldots\ldots\ldots\ldots\ldots\ldots\ldots (7)$$

and if we have a definite slope which we call s. (say $1\frac{1}{2}$ to 1 for earth, or $\frac{1}{4}$ to 1 for rock), we will have

$$H+H_2+h+h_2=W+\frac{W_2-2b}{s} \quad \text{.........................} \quad (8)$$

and the volume equation becomes

$$=\frac{1}{6} L \left\{ W\left(d+\frac{d_2}{2}\right)+ W_2\left(d_2+\frac{d}{2}\right)+\frac{3}{4}\frac{b}{s}\right.$$

$$\left.(W+W_2-2b)\right\} \quad \text{...} \quad (9)$$

from which it will be seen that the volume may be obtained by having the slope stake distances of the two end sections (W and W_2), the centre cuts (d and d_2) the roadbed width (b), and slopes (s), thus eliminating the determination of the middle area from the calculation.

(2) Mean areas, which custom has established as the one to be ordinarily used, because of its simplicity, is merely, volume $= L\frac{A+A_2}{2}$ ....(10)

where A and A_2 are the two end areas, the error involved in this formula is $+\frac{1}{6}L(h-h_2)^2 s$. Where h and h_2 are the two centre heights, and s = slope of earth work, it is evidently = zero when $h=h_2$ and increases with their difference.

(3) Middle area volume is given by

Volume $= L \times A_1$ ....(11)

Where A_1 = area of section midway between the ends of the prismoid.

The error involved in this formula is $+\frac{1}{12}L(h-h_2)^2 s$, being one-half as much as in the formula for mean areas, and also disappearing, when $h=h_2$—and although equation (11) is more accurate than equation (10), it is not used, because once the middle area has been determined, the prismoidal formula is easily applied and still more accurate. There is another reason why formula (10) is not objectionable, that, in general, the profile of a line is convex in cuts, and concave in fills, and any system of sections, no matter how carefully taken will, as an average, need a little allowance for this rotundity between sections.

At points where the cut changes to a fill there should be two grade pegs determined (see Fig. 76) and cross-sections taken at these points, this makes the first volumes in cut and fill always pyramidal, there is no necessity for a centre grade peg, unless the distance from one grade peg to the other is excessive.

Borrow pits should be carefully cross-sectioned from some well defined base line before excavation is permitted, and, if at all possible, it should be made imperative that these pits before being abandoned should be left in good shape for final cross-sections and for drainage; undrained borrow pits are unsightly, a menace to health, and difficult, often, of measurement. There is another matter in this connection which should be well attended to, *i.e.*, proper referencing of alignment hubs; this may be done by cross lines fixed by hubs, trees, etc., or by right-angled lines and steel tape measurements, but in whatever way accomplished, considerable judgment is required to place the references out of harm's way and at the same time reasonably available; in side-hill cross-sections made after excavation great precision is required in this respect to prevent serious error.

ARTICLE 29.—METHODS OF PAYMENT AND CLASSIFICATION OF MATERIALS, ETC.

There are occasional instances of railway companies of some financial strength and progressive growth carrying on construction under their own management, with their own plant and by day labor, but such instances are not frequent, and in general we may look to responsible contractors for the rapid execution of this kind of work requiring experience, undertaking risk, and with considerable capital as plant continually wearing out. Occasionally such work is taken by contractors at so much per mile, within limits of curvature, grades, style and locality, but as the element of risk is great the price is correspondingly high. Only approximate estimates are furnished, and the experienced judgment of the contractor to size up the class of material to be met with is his chief reliance. Such contracts are apt to be made when the railway company and the contractors are more or less identical.

Again, at times, contracts are taken in which such as timber, stone and iron are specified as to quality and price *pro rata*, but the excavation is unclassified and an average price per cubic yard is given, the contractor again taking chances, and being, by this method, unable to alter the location, his risk is great and price high enough, on the whole, to cover the risk; this has led up to what is, at present, the general method of letting contracts for excavation, namely, to define certain classes of material as rigidly as possible and fix prices for each class, the usual divisions are solid rock, loose rock, hard-pan and other cemented material, and earth. As the dividing lines between these classes are purely arbitrary they need to be defined for all possible contingencies, which is a difficult matter. An engineer is always, although in the employ of the railway company, more or less an arbitrator, and he should endeavor to be just to all; theoretically he should always live up to the strict letter of the contract, and ordinarily this is the only course to pursue, but there are cases in which contractors, in their eagerness to obtain contracts, take them at too low prices, or they may strike some very difficult cuts which will not classify very highly if the specifications are adhered to, and in such cases it is usual to allow percentage classifications based on a fair cost of doing the work, *e.g.*, a heavy cutting composed of a mass of small boulders closely cemented together would only classify as hard-pan, but often must be excavated entirely by drilling and blasting, and in such a case percentages, at least, of rock would be quite justifiable.

This idea of *helping out* a contractor, however, is a very pernicious one, and should only be done for good cause, where the recipient is worthy of it by his economical handling of the work, and with full knowledge and consent of the railway company. The vigilant watch and full knowledge of the various classes of material met with in excavation and the most economical methods of handling them, form one of the most important duties that a railway engineer has to perform, needing, moreover, a knowledge of men as well as ways and means.

The calculation of quantities in structures should be made very minutely and in detail, as the prices are so much higher per unit than those of excavation; but the method to be used will depend on the terms of the contract and the individuality of the engineer. In some cases payment is made on bills of timber furnished, and on general plans of masonry, while in others the actual timber used and the masonry as built are the basis of payment. In the case of timber this latter method should include payment, as timber delivered, for all pieces cut off either from piles or timber, so long as the material used was as per bill given.

Quantities of earthwork should always be measured in excavation, for no one can determine accurately the shrinkage of fills, especially as only a portion of it takes place while construction is in progress, and continues for a year or two depending on the method of forming the bank. The total shrinkage is fairly well known, being about 5 per cent. for sand, 10 per cent. for clay, and 15 or 20 per cent. for loam; while rock expands 50 to 75 per cent., depending on the size of the pieces, and such figures modified according to the age of the bank will be sufficiently accurate for monthly estimates, but not for final ones. It is more laborious, often, to measure irregular borrow pits, but by insisting on these pits being shaped up before being left, the extra labor is not so very great and is the only really reliable way, the chief value of embankment quantities being to aid in a proper distribution of material from cuts, to enable overhaul calculations being made, and for approximate estimates where errors of 5 per cent. or even more are not objectionable, as the company, in its monthly payments to contractors, reserves 10 or 15 per cent. for just such exigencies. In taking monthly estimates, the only certain method is to make each one a total estimate in fact as well as in name, and derive the current estimate by deducting the total one for the month previous from it, that is, never take notes only of what is thought to have been done during the month, for nothing is more difficult or more apt to lead to error—whereas if the total work done or material delivered at each point is noted, the errors of each month are eliminated in the next one—the extra labor involved in this is usually insignificant.

Specification for Excavation.—Materials excavated will be classified as earth, quicksand or dry hard-pan, loose rock and solid rock. Earth includes everything except the other classes mentioned. Hard-pan includes all cemented clay or gravel, or any combination of these that it is not practicable to plow with a four-horse team. Loose rock includes all stone containing not less than one cubic foot, and masses of detached rock containing not over one cubic yard; also all slate, shale or other soft rock which can be removed without blasting, although blasting may be resorted to. Solid rock includes all loose rocks containing over one cubic yard, and all rock in place which requires drilling and blasting.

(N.B—.The earth and hard-pan are sometimes placed in one class as earth, at a higher rate, thus saving much contention.)

ARTICLE 30.—SURFACE DRAINAGE.

In addition to the provision for flow of water under or through the track, there is yet the question of track and slope protection which is of almost equal importance. Where, on side-hills, the surface flow toward the top of the cut slopes and toe of the embankment slopes is considerable, ample provision should be made to intercept it. In general, catch water ditches three or four feet wide, and one to one and one-half feet deep, should be dug so as to run in a continuous line along the upper side of the cuttings and embankments from each lateral watershed to the nearest culvert or stream—which should set back five or six feet, generally, from the edge of the slopes leaving a solid berm; the material from these ditches, cast on the lower side will form an additional protection. In very porous soils, these ditches may need to be lined with pitch or planked; but in any case will prevent that heavy washing down of cut slopes which, otherwise, fills up the track ditches and floats the track, making the roadbed soft.

The track (cut) ditches themselves should be turned into these catchwater ditches at the upper grade points, so as not to empty down along the toe of the embankment, eating it away; and in very wet cuttings, it may be necessary to run a farm tile about three feet under the track ditches and parallel to them to aid the drainage. In case of quicksand, which will soon fill up tiles, the tile must be covered with straw and laid with collars, or longitudinal round pole drains may be substituted for the tiles.

The slopes of cuttings themselves also need protection to prevent erosion ; in ordinary cases the sowing of grass seed on the slopes of cuts and fills will answer the purpose, but where the cut slopes are wet and springy it may be necessary, in addition, to cut a series of diagonal ditches on the slopes to bring the water to the cut ditches by easy grades ; in extreme cases the construction of a network of tiles on the slopes may be necessary to effect complete protection. Perhaps the most imperative matter of all is to have the ordinary cut ditches always cleaned out—free from boulders, ties, ballast, etc., which tend to accumulate during maintenance. The form which such ditches assume is of a wedge shape, with a slope from the track of about 3 to 1 and an outer slope the continuation of the cut slope. This form will tend to maintain itself better than one with a flat bottom and steeper slopes.

CHAPTER VI.

RAILWAY LAW.

Railroads being recognized as public necessities have had great powers conferred on them by legislatures, which have also necessitated many legal restrictions to prevent the abuse of these powers. All of which in Canada has in time become formulated in the "Railway Act." This Act defines, amongst other things necessary for a railway engineer to be familiar with :—

I. The powers conferred on the Railway Committee of the Privy Council for regulating traffic, tolls, returns, methods of operation, of construction, capital stock, and distribution of gross revenue obtained.

II. The privileges and powers granted to railway companies.

III. The duties of a railway company to the Government and to the Privy Council.

IV. The duties of the railway company to the individual, and the rights of the private individual.

Much of the matter contained is rarely needed by the engineer, and the following extracts cover the main information which he is likely to need in the course of construction and maintenance.

I.—POWERS CONFERRED ON THE RAILWAY COMMITTEE.

(*a*) To regulate the speed through various classes of cities, towns and villages—which is not to exceed six miles per hour in any case.

(*b*) To regulate the use of steam whistles in towns cities, etc.

(*c*) To regulate the means for passing from one car to another, for the safety of employees, and the methods of coupling cars.

(*d*) To impose fines for offences under these clauses.

(*e*) To enquire into, hear and determine applications, disputes or complaints regarding right of way and location questions, constructing branch lines, the crossing of one railway company's tracks by those of another railway company, the construction of railways along or across highways or navigable waters, tolls, rates, running powers, traffic arrangements, unjust preferences, discriminations, distortions and the carrying of highways, streets, ditches, sewers, etc., over or across the lands of a railway company.

(*f*) By itself or agents, it has full legal power to enter on to the property of a railway company to examine books, plans, etc., to summon witnesses and in general it has the same powers as a law court.

(*g*) The inspecting engineers of the Privy Council are to have every desired information, in reason, supplied to them on demand. They are to be carried free while on inspection trips, and to have the services of all company telegraph operators free while on Government business—penalties for obstruction of the inspecting engineers are also defined.

II.—THE PRIVILEGES AND POWERS OF A RAILWAY COMPANY.

These are given with the view of assisting the company in overcoming any obstructive measures of a corporation or individual, where it is evident that the public would be best served by the construction and operation of a railway.

These powers should be thoroughly considered by a company's engineer before taking any steps likely to incur the ill-will of the public, to whom the company must ultimately look for its income.

(*a*) The company or its agent may enter on crown lands or the lands of any person or corporation whatever for the purpose of survey and location.

(*b*) It may purchase land for the use of the railway, and may sell what it does not need.

(*c*) It may build anywhere within one mile of the first located filed line, or within any further distance prescribed by the special Act.

(In the Act a railway is said to be near to another when some part of one is within one mile of some part of the other railway.)

(*d*) It may fell trees within 99 feet of either side of the railway, when they are liable to fall across the track.

(*e*) It may cross or join any other railway, and enter on its lands.

(*f*) It may divert temporarily or permanently streams, highways, water or gas pipes, sewers, drains or telegraph or telephone poles, but must restore them to their former state, if possible, or put them in a state not materially altering their usefulness. (Note this with reference to keeping them usable during the construction of the roadbed).

(*g*) It may construct, operate and keep in repair its road, together with the accessories commonly belonging to a railway, and may carry traffic and collect tolls, and may, in general, do everything necessary to a successful operation of its enterprise and the accommodation of the public, but in the exercise of these or other powers the company shall do as little damage as possible, and make full compensation for damage done or loss inflicted, in a manner prescribed in the Railway or a special Act.

The essence of these general powers is compensation. It is the limitation of what would otherwise be absolute, arbitrary powers.

II.—(*a*) POWERS WITH LIMITATIONS.

(*a*) No person who holds a contract for work with a railway company can be a director on its board, nor can a director or officer of a company even go surety for a contractor.

(*b*) The first charges on the income of a railway company are the penalties, if any, arising from this Act. The next are the working expenses of the road, and the third are the bonds—the latter, however, all having equal claims in proportion to remaining assets.

(*c*) The consent of the Governors-in-Council is necessary before Crown or Indian lands can be entered on, and in the case of Crown lands the right of way only, i.e., an easement, is all that can be obtained, and not ownership or full possession.

(*d*) The consent and approval of the Railway Committee must be obtained before possession or use of the land or property of another railway company can be effected.

(*e*) The ordinary amount of land obtainable without the consent of the owner is 99 feet, while exceptions are authorized by the Minister in case of deep cuts or fills and depot grounds, which is in general limited to a tract of land 1,950 feet long by 300 feet wide, but any extent of land may be purchased with the consent of the owner. Extra land is to be shown on the maps or plans filed with the Government.

(*f*) After filing plans (to be afterwards explained) the company may take possession as shown on those plans, and settle afterwards, amicably or by arbitration. This is a very necessary power, for otherwise contractors would be kept off the land for indefinite periods, the progress of the work impeded and a good chance given the contractors for suits for damages caused by delay, or for excuses for slow progress.

(*g*) The company may enter on lands not more than 600 feet from the located line, for purposes of construction or repairs, without consent of the owner, provided a sum of money, fixed by a judge of the Superior Court, is deposited with that court, pending the award for damages.

(*h*) Power of surveying and arbitrating in the usual way is also given whenever the company desires extra land for stone, gravel, water or earth, or on which to construct sidings or branches, or on which to convey water to the company's works.

(*i*) The company may occupy land between 1st November and 1st April, with snow fences, subject always to damages as a court may decide.

(*j*) No lateral deviation of more than one mile shall be made from the original location, except under provisions of a special Act.

(*k*) Error in name or omission of the same from plans and books of reference does not prevent the company from entering on land so affected.

III.—DUTIES OF A RAILWAY COMPANY TO THE GOVERNMENT.

(*a*) On completion of location surveys, plans and books of reference showing all properties asked for, must be forwarded in triplicate to the Minister, who after ten days' notice to all interested parties, hears all counter claims and representations, and the plans, as finally decided on, are signed by the Minister, after which one copy is retained at the department, another is deposited by the company in the office of each municipality affected, while a third one is given into the hands of the company itself. After these proceedings following the action as described in Art. *f*, powers with limitations.

(*b*) All extra widths desired must be shown on the plans filed.

(*c*) Any change from the original plans or profiles must have separate plans, etc., submitted and deposited in the usual way before the alteration can be made.

(*d*) Ten days after plans have been filed in the offices of the municipalities (registry offices) and notice published in a newspaper, a notice may be served on the interested parties owning land, giving

1. Description of land required.

2. Amount offered by the company for land or damages.

3. Name of company's arbitrator if the offer is not accepted.

Such notice to be accompanied by a sworn statement of a surveyor or engineer (not an arbitrator) stating

1. That the land is needed as described for the purposes of the railway.

2. That he knows the land, or damages likely to result, from the railway being built and operated.

3. That the sum offered is, in his opinion, fair compensation.

(*e*) Should the offer not be accepted, arbitration is resorted to. This is usually outside the province of the engineer, except in giving evidence and preparing plans, but it may be useful to remember that, in considering the value of land taken or amount of damages done, the increase of values of lands adjacent (*i. e.*, of same plot) not taken by the company, which is created by the construction of the railway, is to be taken into account and offset against any damage inflicted.

(*f*) Before a company can unite with or cross over any railway with its roadbed or track, it must submit plans and full details of proposed mode of crossing to the Railway Committee, and give ten days' notice of application to the other company affected. The committee may make such changes or regulations as appear necessary for public safety, and apportion the cost of constructing the necessary works to the different companies, and may also, in case of level crossings, on application of either company, direct such inter-locking signal system or device to be used as, in their opinion, renders it safe for engines to pass over such crossing without coming to a full stop. This clause also applies to electric and other street railways.

(*g*) When the railway is finished, plans and profiles in triplicate of the completed work, showing land taken, names of owners, etc., must be filed as in the original survey or alterations within six months after completion of the railway, under penalty of $200 fine per month.

(*h*) The scales of plans and profiles are to be as prescribed by the Minister; they are usually: Plans, 400 feet to one inch; profiles, 400 feet to one inch horizontal, and 20 feet to one inch vertical, and are to be all signed by the chief engineer or president of the railway.

(*i*) Within 48 hours at furthest, after any accident has occurred on a railway line, involving personal injury, or the damage or destruction of any structure, the railway company must notify the Minister of the same under $200 per day penalty. And, if necessary, the Privy Council may appoint a commissioner to enquire into the causes, etc., of the same.

(*j*) Annual returns from 1st July to 1st July must be forwarded in duplicate to the Minister within three months after the expiration of such financial year by each railway company, of its capital, traffic, working expenditure, and any other information shown on the blank forms furnished.

(*k*) Weekly returns of traffic shall also be furnished the Government within one month after the period quoted, and a copy of the same must be posted for public view in the head office of the company.

(*l*) Twice a year accidents and casualties must be similarly reported within one month after each six months' period has elapsed, giving, (1) Causes and natures, (2) locality and time of day, (3) extent and particulars.

IV.—GENERAL DUTIES OF A RAILWAY COMPANY TO THE PEOPLE AND THE RIGHTS OF INDIVIDUALS.

(*a*) A company may not obstruct the entrance to a mine, open or about to be opened.

(*b*) A company shall not impede navigation in a navigable river.

(*c*) In constructing drawbridges, bridge-piers or wharves, the manner of construction as it affects navigable waters shall be fully and entirely decided upon and directed by the Railway Committee, under heavy penalties for disobedience.

(*d*) Highway crossings.

I. A railway shall not be carried along a highway, but shall merely cross it, unless permission has been obtained from the Railway Committee, and, in crossing a highway, a good safe passage for vehicles must be kept open continually and no obstruction offered to travel.

II. In a level crossing, the surface of the rail must not be more than one inch above or below the general surface of the crossing.

III. When a highway passes under a railway, it must have at least 20 feet clear width and 12 feet clear height, and the gradient of the highway shall not exceed one in twenty, unless originally greater and left undisturbed.

IV. When a highway passes over a railway, the approaches shall not have a gradient of more than one in twenty, and must be fenced at least four feet high, and the bridges by which these highways are carried over the railway shall have a clear height of seven feet above the highest freight car hauled over the road, and over 14 feet clear width shall be given on approaches and bridge.

V. The company shall present to the Railway Committee (which notifies the municipality interested in order that they may oppose by a delegation) a plan and profile of every proposed highway crossing, and the committee will then decide whether it is to be changed in any way, or if a level crossing is approved of, whether a gate and watchmen are necessary to public safety, and the decision of the committee regarding details of construction, and also as to apportionment of costs between the company and other parties interested, is to be final, and followed out within a prescribed time under penalty.

VI. Sign-boards with letters six inches high shall be placed at every highway level crossing.

(*e*) Farm crossings.

One, at least, shall be made for every separate portion of land and for each disconnected portion thereof, so located as to be difficult of access otherwise than by means of a crossing of the railway.

(*f*) Bridges and tunnels.

Shall be built, maintained (and raised if necessary, whenever repaired or reconstructed), so as to maintain a clear height of seven feet between the top of the highest freight car used and lowest beam or obstruction of the bridge or tunnel.

N. B.—The Governor-in-Council may except from this clause any railway on which air brakes are used exclusively. No company shall run cars over any bridge unless constructed and maintained with safeguards, and of strength approved of by the Minister.

(*g*) Fences and cattle-guards.

Whenever municipalities are surveyed or settlement exists, fences and cattle-guards shall be built and maintained, and in case of adjacent land being occupied, this must be done as fast as rails are laid, and the company shall provide gates having proper fastenings or hurdles at all farm crossings. When this is done, it is the duty of the landowner to keep the gates closed when not in use. If left open accidentally, no action for damages against the company can be sustained, and if left open purposely, a counter-claim for damage may be entered in addition to inability of recovering for loss of animals, etc., resulting from such leaving open of gates, etc. (Fences must be turned in to the cattle-guards).

(*h*) Opening of the railway for traffic.

No company shall open its road for passenger traffic until one month after giving notice in writing to the Minister, and after an inspecting engineer sent by the Government shall have reported favorably as to the safety of the road, strength of structures, and adequacy of rolling stock.

(*i*) Repairs.

Upon complaint of the officers of any municipality, the Government shall send an inspecting engineer to examine the condition of the road, and the company shall at once make such repairs as he considers necessary for public safety. Until such repairs are made the engineer may limit the number, speed or weight of trains and engines passing the point under repair.

(*j*) No discrimination in tolls between different persons or companies for the same service shall be allowed; any special rate allowed to one must be allowed to all.

(*k*) No secret special rebate or toll shall be allowed, and any company shall on demand make known to anyone any special rate, toll or rebate.

(*l*) No discrimination between places shall be allowed, unless a lesser rate is necessary to secure freight at a competing point. (This, therefore, protects local points only against one another, and not against railway centres).

(*m*) Every company shall afford full and equal advantages to all persons wishing to ship or travel over their railway, and shall afford equal facilities to each and every connecting line or lines of railway for transfer of traffic.

(*n*) Every company affording facilities to any express company shall grant the same to any other express company demanding it.

(*o*) Every agent of the company shall receive freight or any allowable traffic when offered for shipment, under penalty.

(*p*) Every train before crossing a draw or swing bridge shall stop at least one minute to ascertain from the bridge tender that the bridge is closed and safe to cross.

(*q*) At least 80 rods before crossing every highway (on level) a bell must be rung or whistle sounded, and this must be continued at short intervals until the crossing is passed.

(*r*) An officer shall be stationed at every level crossing of two railways, and no train shall pass over it until a signal has been made to the conductor that the way is clear.

(*s*) Every train shall stop one minute at a level railway crossing as in No. 16, unless there is an interlocking system, when they may pass at such speed as the committee may allow.

(*t*) No train shall pass through thickly populated towns, etc., at more than six miles per hour, unless the track is fenced.

(*u*) No train or car shall be allowed to stand on a highway crossing more than five minutes at a time.

(*v*) All frogs, wing rails, guard rails, etc., shall be packed up to the underside of the rail wherever less than five inches space exists.

(*w*) Sections (*p*), (*r*) and (*s*) have been recently modified so as to permit of interlocking signals being introduced at junctions and railway crossings, in which case the Privy Council may permit trains to pass in or across at specified rates of speed without stopping, whenever the signals give the right to do so, but if the signals are not satisfactorily worked the council may revoke the permission.

PART II—CHAPTER I.

TRACK.

ARTICLE I.—FORM OF ROADBED.

The first essential of a good track is proper drainage; there can hardly be good track without it, from which it naturally follows that too much care cannot be taken in forming a roadbed at the completion of its construction, which will have good drainage in itself; even with abundant and good ballast, drainage is necessary, while it may be the saving feature of a track surfaced with inferior or scanty ballast. Plate XXII. shows types of road-beds in use in America, and it will be seen that most of them have a slight slope each way from the centre, forming a rounded surface onto which the ballast is laid; the crown at sub-grade should be 3 to 4 inches for a single track in cutting, but may be partially omitted on embankments, as future settlement tends to round off the corners and aid drainage. Should low spots exist in the centre of the roadbed beneath the ballast, water will lodge there and soften up the earth so that the ties will sink under the churning action of car and engine wheels. Although not essential or always done, it is an advantage and an economy of ballast to elevate the roadbed on curves parallel to the expected plane of the ties and rails; this practice also gives an elevated track before ballasting is commenced. Widths of roadbed vary with the climate and materials. Embankments vary from 10 feet for cheaply built roads in the Southern U.S.A. to an ordinary standard of 16 feet for Canadian and Northern U S.A. first-class roads; cuttings vary similarly; but are usually about 6 feet wider than the embankments for making ditches; for purposes of handling snow it is not found advisable to make cuttings less than 22 feet in Canada, although rock cuts with narrow ditches are sometimes made 20 feet. To all of these, 12 to 14 feet are added for each additional track, and in case of very wet cuttings extra width may be needed for proper drainage (see Fig. 8, Plate XXII.), or a tile may be laid beneath the cut ditches to drain the sub-soil (see Fig. 5, Plate XXII.). Ordinary cut ditches are about three feet wide and one foot deep, and may be wedge-shaped (Fig. 7) or trough-shaped (Fig. 8), but although the latter is often dug in the first place, the weight of evidence is in favor of the former, which is formed by a flat slope of from 2 to 1 to 6 to 1, starting from near the edge of the ballast and meeting the cut slope at an angle. The tendency of such a ditch is to direct the water well away from the track and thus prevent undermining of the ballast. Cut ditches should be led well away from the mouth of the cuttings to avoid scouring the foot of the adjacent bank, indeed, the cut ditch on the upper side should join the catchwater ditch and continue down to the entrance of the nearest culvert as a berme ditch placed five or six feet away from the foot of the bank. By a thorough system of ditching at the conclusion of construction much trouble and expense can be avoided and the energies of the track gangs during early maintenance may then be devoted to other things. To make the ditching system complete, catchwater ditches should be dug along the upper side of every cut, placed six or eight feet back from the top of the slope, the earth from them being placed inside; these ditches should collect all those small trickling streams and general hillside wash that would otherwise run down the cut slopes, carrying sediment into the cut ditches. These cut ditches are often soon neglected during early maintenance, and extra ties, heaps of unused ballast and stray boulders block the drainage, while in later years rotten ties and weeds need watching. Too great stress cannot be laid on having clean, straight cut ditches with a uniform fall.

Of late years construction has been usually very rapid, and embankments, if made of earth, will rarely have completed more than half of their shrinkage; this will vary in amount with the method used in building the bank, being greatest when built with wheelbarrows or machine graders from side ditches, and least when flat or wheel-scraper

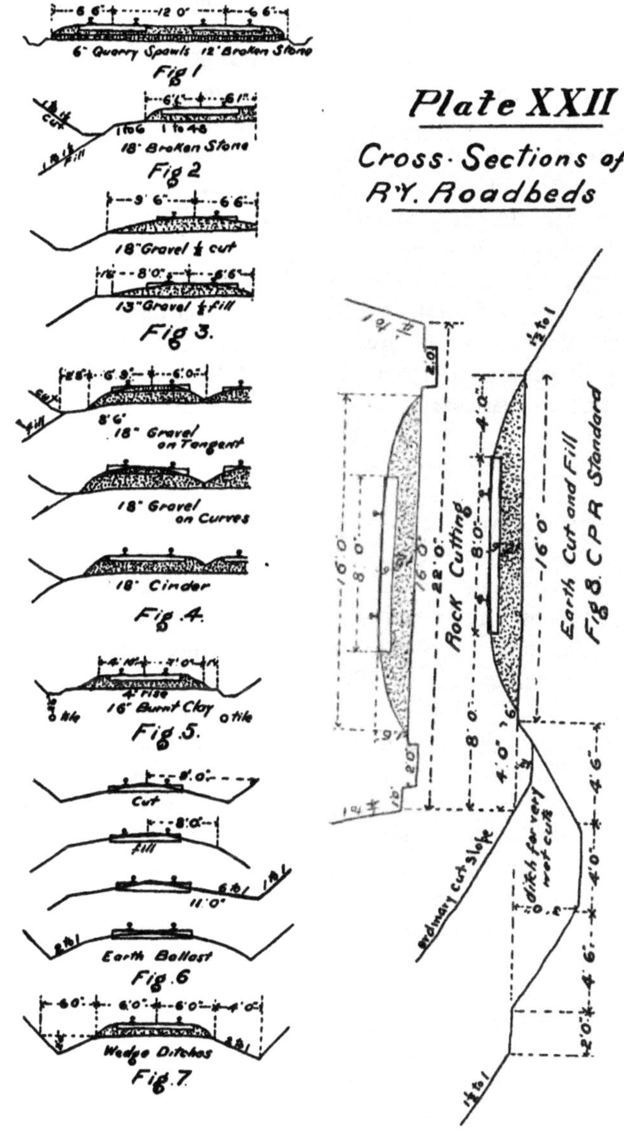
Plate XXII
Cross-Sections of
R'Y. Roadbeds
6" Quarry Spawls 12" Broken Stone
Fig 1
18" Broken Stone
Fig 2
18" Gravel ½ cut
13" Gravel ½ fill
Fig 3.
18" Gravel on Tangent
18" Gravel on Curves
18" Cinder
Fig 4.
16" Burnt Clay
Fig 5.
Cut
Fill
Earth Ballast
Fig 6
Wedge Ditches
Fig 7
Rock Cutting
Earth Cut and Fill
Fig 8. C P R Standard
ordinary cut slope
ditch for very wet cuts

work has trampled it in thin layers by the horses' feet, etc. For these reasons all banks made of earth ought to be left full width and a certain per cent. of height at each point above the theoretical grade line. Of course, abrupt changes in track surface are not desirable, even for a short time, and such allowance for shrinkage should be made with judgment according to the merits of the case in hand at each point, bearing in mind what the ballast is costing, the expense of re-lifting sunken track and the large amounts of extra high-priced material needed if this allowance for shrinkage is not made before grading is completed ; on page 150 the per cent. of shrinkage of different materials is given, which will serve as a basis for estimating how much extra height should be given to the banks ; if construction has been completed in one season at least one-half these amounts are necessary.

ARTICLE 2—BALLAST.

The quantity of ballast used is a purely financial question, and up to a usual limit of 12 inches underneath the ties, the more the better can track be maintained for the same cost ; 12 inches under ties takes about 3,000 cubic yards per mile; 6 inches under ties takes about 1,800 cubic yards, including filling around ties as in Fig. 8, Plate XXII., but with no allowance for sunken banks and extra material. The functions of ballast are :

(1) To afford lateral, longitudinal and vertical support to the ties sufficient to keep the track in line and surface without incessant track labor.

(2) To carry off all water as rapidly and thoroughly as possible after rain storms or thaws.

(3) By drainage to lessen the action of frost in heaving track during the winter and spring.

(4) To give elasticity to the roadbed.

The following materials are used more or less extensively for ballasting and are given in order of merit as nearly as may be :

(1) Broken stone to a 2-inch ring ; coarser underneath.

(2) Furnace slag and cinder.

(3) Coarse, clean gravel.

(4) Broken bricks or any form of very hard burnt clay.

(5) Sand not so light as to be easily blown away.

(6) Earth, usually compact clay, seldom loam.

Broken stone ballast, although expensive and hard to tamp and surface with, gives the most durable and satisfactory track with least labor for maintenance; only roads with heavy traffic can afford to have it, as it costs from 75 cents to $1.25 per cubic yard in place. When used, it is generally flush with the top of the ties for about 1 foot beyond their ends, thus giving lateral support, and side slopes rather steep (about 1 to 1). A very finished appearance can be given by laying a margin of stones to line by hand, and keeping the rest of the roadbed, outside, free of ballast and grass. The slag from blast furnaces, if properly cooled and broken, makes a very good and durable ballast, but its use is evidently limited in area, and the price will vary according to circumstances; cinder also is a valuable ballast, but limited in quantities. Probably gravel may be looked on as the ballast more generally used in America than all other forms combined, because of its wide distribution and general utility. When clean and fairly free from sand and large boulders, it drains well, surfaces easily, and holds track from all but lateral movement; in this it is deficient as it will not stand steep enough to admit of the ends of the ties being fully submerged, unless a very wide roadbed is used. (See Figs. 3, 7 and 8, Plate XXII.). The cost of gravel ballast in place varying with length of haul, may be put at 15 cents to 20 cents per cubic yard if loaded with steam shovels from a good pit and unloaded by ploughs, but will run as high as 40 cents when material is manually handled from pits with heavy stripping. In all cases the stripping of pits should be attended to, and all inferior material wasted or put on low or narrow banks. The ballast material should be of a uniform quality, as any patches of loam or clay mean just so many sunken spots in the track.

Sand ballast creates dust in summer which injures the rolling stock, does not hold a track well to surface or in line under heavy traffic, and has a tendency to hold water and heave track in the spring; unless very coarse it is not at all a good investment if other ballast can be obtained. In such situations many roads have resorted to burnt clay or broken brick, but unless well and uniformly burnt, almost to vitrifaction, it is not a very durable material. In mild climates, such as Southern U.S.A., many railways have ballasted with clay taken from ordinary cuts, either from the cut slopes or hauled by train from the nearest point. If the clay is of a compact nature, and such a cross-section as one of those in Fig. 6, Plate XXII. is used, it will soon get beaten down and shed ordinary rains without any water permeating the roadbed. It is evidently a very cheap way to ballast, and in the absence of other cheap materials may be very justifiably used in such climates by roads of light traffic and meagre resources. Except in the case of broken stone, laid with teams, from adjacent fields, the ballast is put on, after the track is laid, by train loads, and, in so doing, unless the newly laid track is at once roughly surfaced, and trains run very slowly over it until a light "lift" is first put on and the track fairly well lined and surfaced before the ballast trains are allowed to run at a high speed, we may expect permanent injury in the form of bent rails and cracked angle bars, especially as the track is often not fully tied, spiked or bolted. In surfacing and lining track it is well to remember some general principles applicable to all materials and at all times.

(*a*) The coarser material available ought to be put underneath, *i.e.*, on the first lift.

(*b*) When the supply of ballast is limited and sub-grade sunken on the banks, it is better to be satisfied with a track having local depressions below the theoretical grade line, rather than to rob the sides by building up a high, narrow track to the true grade, as such a track will soon sink and get out of line—being deficient in lateral support.

(*c*) Each tie should be tamped equally well, because even one tie, without support, acts like a force pump ; each passing truck, by suddenly depressing it, compresses the air under it, forces out more ballast, until there is a cavity formed, a lodging place for water and a permanent sag in the rail.

(*d*) Ballast should be tamped more firmly under the rails than under the centre of the track, because a centre bearing will cause a rocking motion which will increase rapidly, especially on banks, where the sides are apt to sink more than the centre anyway.

(*e*) Surface is rather more important than alignment, although not so easily obtained or seen by a track foreman.

ARTICLE 3.—TIES.

Ultimately we may expect metal ties to take the place of wooden ones. In Europe, with dear wood and heavy traffic, substantial progress has already been made. In America experimental pieces of track have proven satisfactory in cheapening maintenance, and for many reasons, to be enumerated, we may expect progress to be considerable in the near future, but for many years wooden ties will continue, on this continent, to be the rule, and metal ones the exception, although their use constitutes a heavy drain on our forests, which probably amounts to six or seven million ties per year for Canada alone.

Wooden Ties.—Wooden ties are in general use because they are cheap, and simple in use or renewal, and by the use of preservatives their life may be increased considerably. In Belgium and adjacent countries where mild steel ties are in use, wooden ties are being abandoned in favor of steel ones on the following grounds:

(1) That their price will gradually rise owing to the devastation of forests.

(2) The quality of even the best varieties of wood is variable and an unknown factor, being affected by time of felling, place of growth, seasoning, etc.

(3) Preservative methods fail to produce a uniform material for use.

(4) No timber merchant will guarantee ties of wood, while two-year guarantees can be obtained for steel ties.

(5) There is a loss of interest, due to stacking wooden ties for seasoning, whereas steel ties may be in use, legitimately, even before being paid for.

(6) The difficulty of obtaining a good fastening of the rail to wooden ties, and the constant re-spiking necessary.

(7) The selling price of old wooden ties is less than metal ones even in proportion to their first cost. All of these objections are more or less valid, even in America,

but the lasting and holding qualities are most important. Ties are ordinarily 8 ft. to 8 ft. 6 inches long, 6 to 7 inches thick, and 6 to 9 inches wide on top and bottom. They may be hewn or sawed, the former method producing a more durable tie if not hacked too deep before hewing. The top and bottom faces of a tie should be true and parallel planes, all bark being removed, and in sawed ties the removal of sapwood on the sides will add to their durability. They are usually laid 2 feet centres (2,640 per mile). The two ties at an ordinary angle-bar joint being selected as the widest ones near at hand and placed about 18 inches apart, centres, centrally about the joint, giving a suspended joint, but if the long six-bolted 44 inch angle-bars are used, then three ties are placed at a joint 18 inches apart, centres, one at each end and one in the middle; otherwise it is considered best to sort ties into groups of nearly the same width. It is believed that a random mixture of ties of various widths tends to cause poor track, as the narrow ones will sink more than the wider ones.

Ties are made from lignum vitæ, oaks, chestnut, locust, cedar, pine, maple, cherry, red elm, hemlock, tamarac, beech and spruce, being named, roughly, in order of durability in track, without treatment by preservatives. The life of a wooden tie in track, untreated, varies from 4 to 6 years for the poorer kinds, up to 10 or 15 years for the more durable ones, except lignum vitæ, which lasts 30 or 40 years. The length of life will depend on locality of growth, the kind and amount of ballast used, drainage, amount and speed of traffic, whether the tie is on a curve or tangent, and finally whether the rail rests directly on the tie or on a tie-plate or metal chair of some form. The wear on curves is greater than on tangents, due to the cutting into the ties of the rail base, which accelerates the rot; also, respiking is more frequent on the former; taking the life of a tie on a tangent as 9 years, one on a 2° curve will last about 8 years, 6° curve 7 years, 15° curve 5 years. Softwood ties can scarcely be used in America, owing to the poor hold of the ordinary dog-spike, which cuts and crushes the fibres of soft woods, while with hard woods the fibres are only squeezed back and are still elastic; but in England, with large metal chairs, soft-wood ties are in general use, and attempts have been made here to use cedar alternately with oak, as they both last well, and the latter will hold the spikes; also attempts have been made to nail oak planks on top of soft-wood ties, dove-tail oak bearing pieces just under the rails, and in other ways dodge the main issue, which is the poor holding power of the spike, but none have been very successful. Metal tie plates such as the one shown on Plate XXV. are now used on heavy traffic roads, sometimes only on curves, and latterly under all the track. These spread the load over a larger surface, and are a great improvement, as they enable a cheaper tie and a deeper rail, relatively to its width of base, to be used. When railway managers in America see the wisdom of adopting wood screws or fang bolts, as in England, for holding the rails in place, a much superior track will be obtained even at a small increase of first cost.

There are several ways of increasing the ordinary life, in track, of a wooden tie, for a tie rots by the solidification of fermented sap, assisted by heat or dampness:

(1) By thorough drainage of the ballast.

(2) By having as little sap as possible in the tie by felling the tree in winter, and subsequent natural, steam, or other form of seasoning.

(3) By charring the surface.

(4) By impregnating the tie with an antiseptic to prevent fermentation.

Such chemicals as sublimate of mercury, sulphate of copper, chloride of zinc, and creosote or oil of tar, serve the purpose more or less successfully, especially the last two, and the last one most particularly. Creosoting is done in a closed receiver, after the tie has been air seasoned two or three months, and trimmed of its sapwood, by exhausting the air to suck out the sap from the pores of the wood. Creosote at 120° F. is then forced in at 10 Atmos. pressure, and after one hour the ties are taken out ready for use; soft woods, which are the only ones usually treated, absorb 7 to 9 lbs. per cubic foot, and the cost of treatment has been reduced from 21 cents in 1879 to 10 cents per

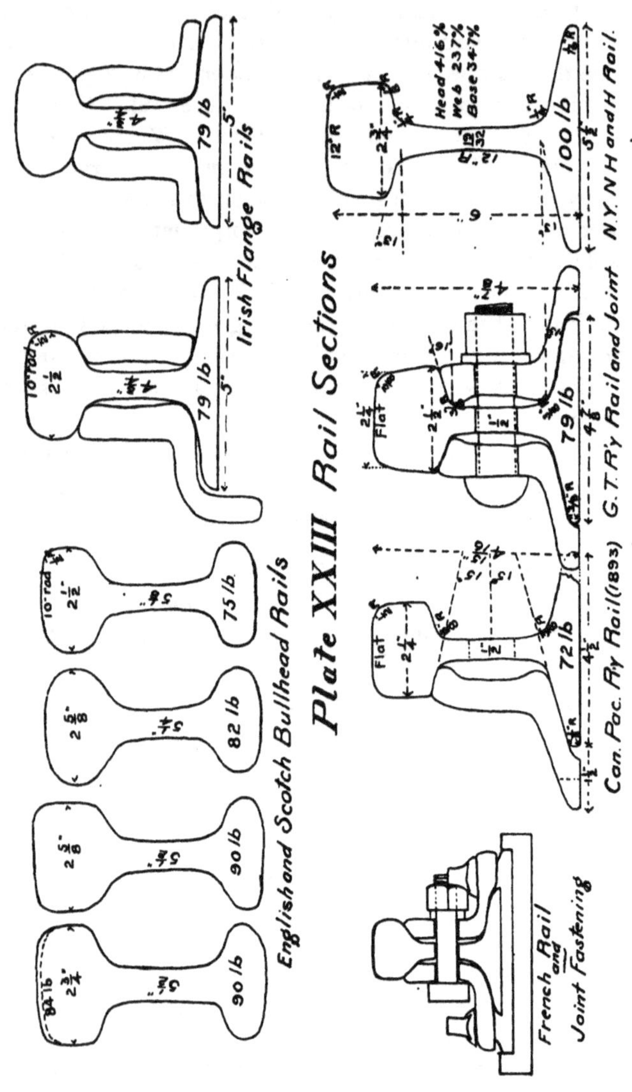
Plate XXIII Rail Sections
English and Scotch Bullhead Rails
90 lb
90 lb
82 lb
75 lb
Irish Flange Rails
79 lb
79 lb
French Rail and Joint Fastening
Can. Pac. Ry Rail (1893)
72 lb
G. T. R'y Rail and Joint
79 lb
N. Y. N. H. and H. Rail
100 lb
Head 41.6%
Web 23.7%
Base 34.7%

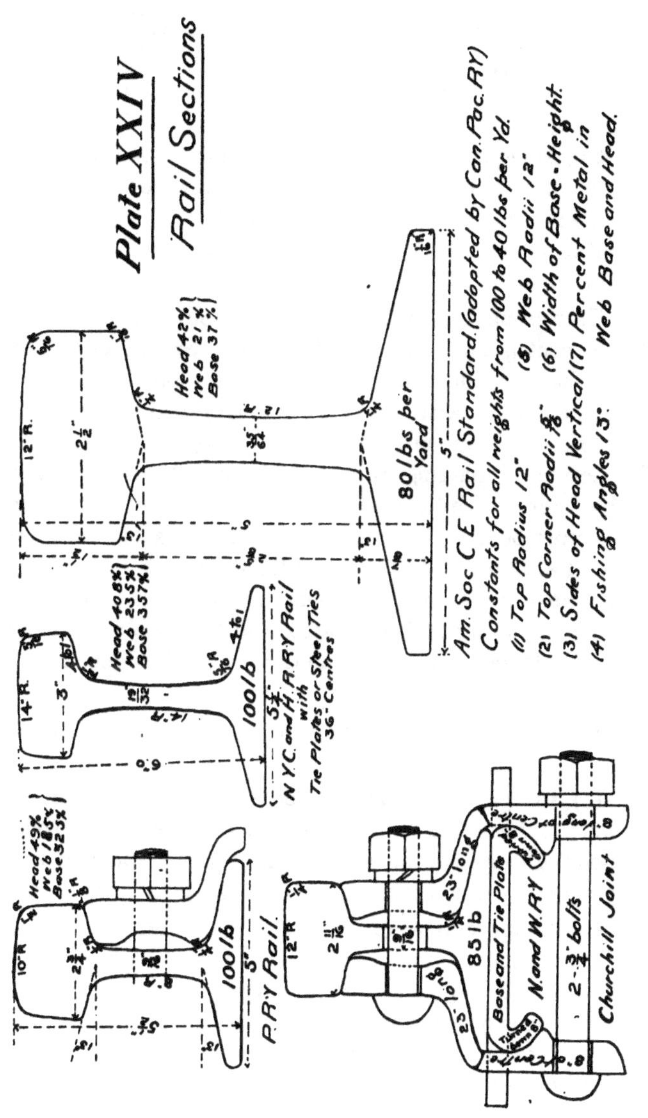
Plate XXIV
Rail Sections
Head 42% Web 21% Base 37%
80 lbs per Yard
Am. Soc C E Rail Standard (adopted by Can. Pac. RY)
Constants for all weights from 100 to 40 lbs per Yd.
(1) Top Radius 12"
(5) Web Radii 12"
(2) Top Corner Radii 5/16"
(6) Width of Base = Height
(3) Sides of Head Vertical
(7) Percent Metal in Web Base and Head.
(4) Fishing Angles 13°
Head 40.8% Web 23.5% Base 35.7%
100 lb
N.Y.C. and H.R.R.Y Rail with Tie Plates or Steel Ties 36" Centres
Head 49% Web 18.5% Base 32.5%
100 lb
P.R.Y Rail.
85 lb
Base and Tie Plate
N and W. R.Y
2-3/4" bolts
Churchill Joint

tie at the present time. The increase in the life of ties in track is greatest amongst soft woods according to the following table:

Timber.	Duration in track.	
	Untreated.	Creosoted.
Oak....................	13	19
Pine....................	7	15
Fir (Spruce)............	5	9
Beech..................	3	16

Creosoted ties will not resist the cutting of rails more, nor are they stronger than untreated ones, but, especially in thickly settled countries, discarded ones will be more valuable as fence posts or fuel, being worth from $\frac{1}{8}$ to $\frac{1}{10}$ of first cost.

Creosoting does not assist ties to hold spikes, and in this respect wooden ties are deficient. Spikes with hard-wood ties on roads of moderate traffic are one thing, with soft-wood ties or with any tie on heavy traffic roads are another. As they are continually being pulled loose by the action of passing trains, and have to be redriven, in the future, with heavier traffic, rails and engines, something must be done to remedy this weakness of American track, the solution of which will lie along two lines, either metal ties and appropriate fastenings, or oak or other durable ties along with tie plates, and fang bolts or wood screws as fastenings—either method will allow deeper rails to be used, or ties spaced farther apart.

Metal Ties.—Three types of metal rail-supports are used:

(1) Longitudinal flanged sleepers giving a continuous support to the rail, and held to gauge transversely by rods; sections of these are shown on Plate XXV. (a) and (b); they have never come into anything like general use.

(2) A succession of cast iron inverted pots, filled inside with ballast and connected, transversely by rods, as in class (1); this method has been used in regions of brackish soils where cast iron rusts less than steel, and can be made heavier, as it is a cheaper material; this method is also only in limited use.

(3) Metal cross ties of inverted trough sections are steadily increasing in favor and are likely to obtain, in the future, general adoption.

The tendency of metal cross-ties is to decrease maintenance charges year by year, while with wooden ones, especially on curves, the reverse is the case. Of these the Post tie seems to be the favorite in Europe; on the Netherlands railway, maintenance with metal ties was about one half of what it was with oak ones, with thirty trains per day and engines of fifty tons, and no ties reported broken. A sketch of this tie is given on Plate XXV. (c); it is of mild steel, weighing 110 to 120 lbs. each, and costing a few years ago $22 to $26 per short ton, with two-year guarantee. It is closed at the ends, narrow and deep at the middle, with thickness varying, being greatest at rail seats; the bottom edges are in the form of ribs ¾ inch thick, projecting ½ inch. The general thickness is ¼ to ½ inch. The narrowing in and deepening at middle gives transverse strength, and prevents the track from creeping longitudinally, or forming a hog back at the centre. The rails are fastened by bolts with T heads and eccentric necks. These bolts pass through the tie from underneath, and into a crab washer which bears on the rail flange and tie; a Verona nut-lock and a nut complete the fastening, and an oblong hole through the ties allows adjustment on curves. This tie presents economy of material and maintenance and general efficiency. It has been in long, extensive use in Belgium, Holland and France, and is probably the best metal tie yet devised for flanged rails. In the United States the Hartford tie has been used with good results on the New York Central, and it appears in general to be an imitation of the Post tie, with an endeavor to simplify manufacture, see (d) Plate XXV. Other forms of less tried qualities are the Standard, an inverted channel beam, and the International, having a section like an elongated bracket, which would appear to be deficient in vertical stiffness. It is probable that persistent attempts at improvement will have a tendency to cheapen manufacture, and hasten the introduction of metal ties on many progressive railways having heavy traffic.

ARTICLE 4.—RAILS.

The progressive history of rails from the first longitudinal wooden sleepers up to the present would be interesting but not in place here. We have arrived at two types, one used in England and Scotland, and in some British colonies and dependencies, etc., i.e, the bullhead or double-headed rail, resting in cast-iron chairs, the other used in the world generally, otherwise, (i.e)., the Viguoles or flanged rail, which is self-supporting.

(A) Plate XXIII. gives sections of bullhead rails, and on Plate XXV. is shown a cast-iron chair for fastening the rail to the ties, and which adds $1,500 to $2,000 per mile to the cost of the track. The original idea involved in the use of this section was to obtain a reversible rail which would double the wearing value if it could be turned over and used again after one head had worn down, but when it was found that the chairs damaged the rail so that they could not be reversed advantageously, this idea was abandoned, and the section now used has a much larger per cent. of metal in the head than in the base of the rail. The British railways use rather heavy rails considering the light rolling stock, but space their ties 2 feet 6 inches apart, centres, due to the superior supporting qualities of the cast-iron chairs; and, in general, the tracks are very solid and first-class, the rails being held to the chair-seats by long tapering oak keys which are tightened occasionally, while the chairs themselves are fastened to the ties with wood screws and bolts, and even those few British or Irish roads which use flanged rails use the same fastenings with tie plates, not trusting to spikes except at every other tie at the most. A special advantage in using rail chairs is that creosoted pine ties become available, and they are probably the most durable and economical tie in use, where it becomes possible to fasten the track securely to them.

(*B*) *Flanged Rails.*—The objections urged against flanged rails, that they cut into the ties, and that they cannot be held properly for heavy traffic with spikes, are overcome by adopting tie plates and screws or bolts for fastenings, and the idea that they are not rigid on curves is shown to be erroneous, as witness the very heavy engines of America running at high speed around much sharper curves than are used in England.

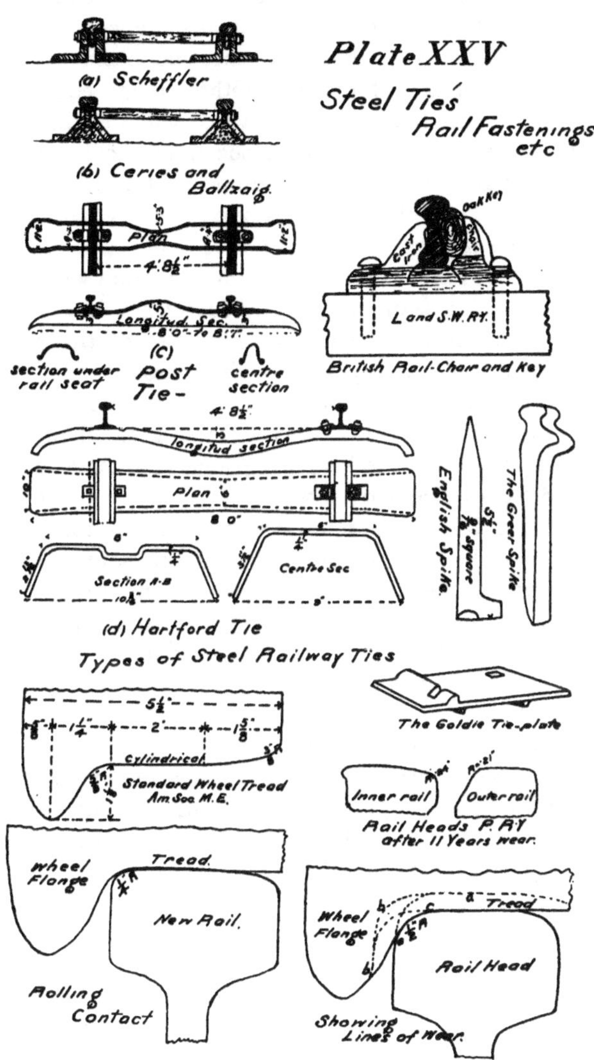
Plate XXV
Steel Ties
Rail Fastenings etc
(a) Scheffler
(b) Ceries and Ballzaig.
Plan
4' 8½"
Longitud. Sec.
8' 0" to 8' 7"
section under rail seat
(c) Post Tie-
centre section
Oak Key
Cast Iron
Chair
L and S.W. R.Y.
British Rail-Chair and Key
4' 8½"
longitud section
Plan
8' 0"
Section A-B
Centre Sec
English Spike.
⅝" Square
The Greer Spike
(d) Hartford Tie
Types of Steel Railway Ties
The Goldie Tie-plate
5½"
2"
cylindrical
Standard Wheel Tread Am. Soc. M.E.
Inner rail
Outer rail
Rail Heads P. R.Y after 11 Years wear.
Wheel Flange
Tread
New Rail.
Rolling Contact
Wheel Flange
Tread
Rail Head
Showing Lines of Wear.

Plates XXIII. and XXIV. give sections of flanged rails of various designs and origins. In detail they will be found to vary widely, but with the exception of the New York Central rail, which has a narrower base for use with tie plates or steel ties, the height is usually equal to the width of base. The first difference noticeable is the per cent. of metal in the head. Other things equal, the more metal in the head the more wear will be obtained, but rails with relatively heavy heads never cool equally, causing initial strains in the section, and a deep heavy head will not get well rolled, and being spongy will wear rapidly when the top layer is gone. The endeavor now is to get a rail as hard as possible, chemically, that will stand drop tests, with a wide, moderately deep head, but not so deep as to induce sponginess in the centre of the head. A wide head is necessary with modern heavy engines to prevent undue crushing of the top surface, due to heavy concentrated wheel loads, and this forces a small proportionate depth of head to keep the per cent. of metal in the rail head from being excessive.

Striking differences in rail design occur in the radius of the top of the head, the upper head corners, and in the side slopes of the head. The tendency in America is toward a flat top, sharp corners, and vertical sides, which is the reverse of English practice of round tops, easy corners and sloping sides, while fishing angles are getting flatter and tend to become standard at 13°.

Plate XXV. gives a standard U. S. wheel tread—rails after eleven years' wear on curves, and two drawings which contrast the fit of a wheel on a rail head of sharp corner radii with that on one of larger radii. It will be seen by the dotted lines that normal wear is upward and outward, thereby increasing the arc of contact between wheel and rail, thus also increasing the resistance and wear, so that the longer this can be deferred by starting with a sharp corner radius and vertical sides, the better, as the contact is then a rolling one only, and the wear and resistance small. Note that the radii of worn rail corners is still about ¼ inch, and investigation has shown that sharp radii of upper corners of rail heads do not cause sharp flanges on wheels, which has been the chief objection raised against them in the past.

Composition of Rails.—When steel began to replace iron as a material for rails it was found necessary to remove the notches in the flanges from the centre to the ends, and even omit them altogether to prevent breakage, the notches being put in the flanges of the angle bars instead, so as to prevent creeping of the track. Rails were made hard to stand wear. Then drop tests were introduced to detect brittleness, and soon forced soft rails to be used, but going to the other extreme the rail heads wore out very quickly, especially as the demand for cheapness produced insufficiently rolled rails. Now there is a gradual tendency to get as hard a rail, chemically, as will just stand the drop tests.

Specifications for Chemical Composition of Rails:

(1) Sandberg (Sweden)—Carbon, if alone, $\frac{5}{10}$ p.c., but only $\frac{3}{10}$ p.c. in presence of $\frac{1}{10}$ p.c. phosphorus; silicon, at least $\frac{1}{10}$ p.c. to give sound ingot and make rail wear.

(2) G. T. R. (Canada)—Carbon, $\frac{4}{10}$ to $\frac{5}{10}$ p.c., sulphur, $\frac{7}{100}$ p.c. or less, phosphorus, $\frac{7}{100}$ or less, silicon, $\frac{1}{10}$ p.c., manganese, $\frac{1}{10}$ p.c.

(3) New York Central Railway (Dudley)—60 to 70 lb. rail: Carbon, $\frac{45}{100}$ to $\frac{55}{100}$ p.c., manganese, $\frac{8}{10}$ to 1 p.c., silicon, $\frac{1}{10}$ to $\frac{15}{100}$ p.c., sulphur, $\frac{7}{100}$ p.c. or less, phosphorus $\frac{6}{100}$ p.c. or less; 70 to 80 lb. rail.: Carbon, $\frac{5}{10}$ to $\frac{6}{10}$ p.c., manganese, $\frac{8}{10}$ to 1 p.c., silicon, $\frac{1}{10}$ to $\frac{15}{100}$ p.c., sulphur, $\frac{7}{100}$ p.c. or less, phosphorus, $\frac{8}{100}$ p.c. or less; 100 lb. rail: Carbon, $\frac{65}{100}$ to $\frac{75}{100}$ p.c., manganese, $\frac{8}{10}$ to 1 p.c., silicon, $\frac{1}{10}$ to $\frac{15}{100}$ p.c., sulphur, $\frac{7}{100}$ p.c. or less, phosphorus, $\frac{8}{100}$ p.c. or less.

Dudley, also regarding different constituents that affect the quality of rails, says: Manganese takes up the oxide of iron, and prevents red shortness, but over 1 p.c. makes rails not only hard but coarsely crystalline, with a tendency to brittleness, flowing easily under wear and oxidizing rapidly in tunnels. Silicon produces solid ingots, free from blow holes in columnar structure, with small compact crystallization. Sulphur causes red shortness and seamy heads; it also tends to check welding of blow holes and ingot pipes. Phosphorus increases the size of crystals and produces brittleness; it must therefore be very low in high carbon rails, which make prices higher, as most ores have phosphorus in them.

Physical Drop Tests for Rails:

(1) Intercolonial Railway of Canada—Supports 3 ft. 6 inches apart; a rail 12 ft. long is to stand one blow of 2,000 lbs. falling 18 ft., and three blows falling 6 feet for 67 lb. rail, with a deflection of 3 to 3½ inches for first, and 2¼ to 3¼ inches for second case. (Drop tests for U. S. roads about the same.)

(2) Irish Flange Rails.—(a) Supports 3 ft. 6 inches apart, a rail not to deflect more than ⅜ inch with permanent set not more than ⅛ inch for 30,000 lbs. at centre for 30 minutes. (b) Same supports, rail to stand 2 blows without breaking, and not to deflect more than 1 inch for 2,000 lbs. falling 8 feet.

Under wear the top surface of a rail head gets more or less cold-rolled and brittle for about 1/32 inch, which is the cause of heads breaking downwards (e.g.) a broken wheel may hammer and cause the brittle layer at top to crack, and the crack will continue on down until the rail breaks. High stiff rails with a broad head are more needed as the wheel loads on drivers get greater, so as to keep a decent track and prevent cold rolling. (Large drivers are not so hard as small ones on track.)

The endeavor is to get a high carbon rail and work it until it is tough and compact in texture in the head.

ARTICLE 5.—RAIL JOINTS.

While great progress has been made in the strength and rigidity of rail joints, they can hardly be considered yet equal to the criterion of simplicity, and of being as strong as the rail itself, and as stiff laterally. Sandberg, by watching the effect of trains on narrow notches cut in the heads of solid rails, concluded that the lipping down was due to lack of support of the fibres, and that we may, therefore, not expect to ever obtain a joint so perfect as to prevent this wear entirely. Various joints are shown on Plate XXVI., and also special ones attached to rails on Plates XXIII. and XXIV. Of these the simple fish plates were considered sufficient in early railroad days, when wheel loads were light and speeds not excessive, but, as these increased, the joints could not be kept in surface, and a lower flange was added, giving us the angle bar,

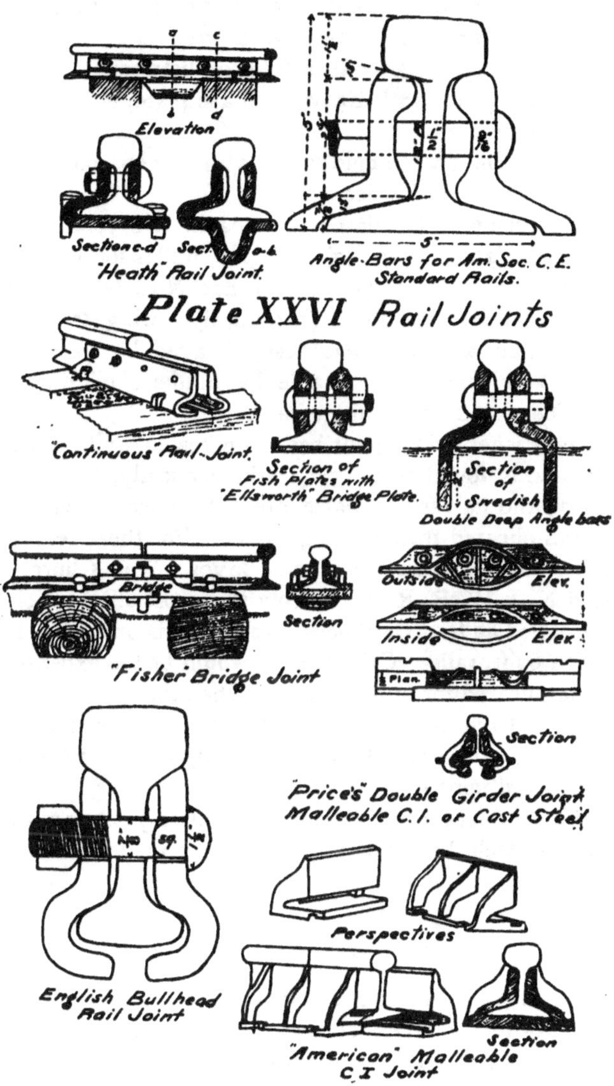
Elevation
Section c-d
Sect. a-b
"Heath" Rail Joint.
Angle-Bars for Am. Soc. C. E. Standard Rails.
Plate XXVI Rail Joints
"Continuous" Rail-Joint.
Section of Fish Plates with "Ellsworth" Bridge Plate.
Section of Swedish Double Deep Angle bars
Bridge
Section
"Fisher" Bridge Joint
Outside Elev.
Inside Elev.
Plan.
Section
"Price's" Double Girder Joint Malleable C. I. or Cast Steel
English Bullhead Rail Joint
Perspectives
Section
"American" Malleable C. I. Joint

which is the ordinary standard form to-day. It is simple, easily attached, etc., and may be used as a suspended joint on two ties with four bolts, or a longer one (44 inches), with 6 bolts, is often used, resting on three ties, and although more expensive, gives better results.

A comparison was made in Sweden between:

(1) Fish plates with Ellsworth base plate.

(2) Angle bars.

(3) Double deep angle bars with 2-inch extension downward between the ties.

The renewals for flattened ends in five years were (1) $6\frac{4}{10}$ p.c., (2) $14\frac{4}{10}$ p.c., (3) $17\frac{6}{10}$ p.c., but as for stiffness they were (1) $\frac{1}{3}$, (2) $\frac{2}{3}$, (3) 1. So that Nos. (2) and (3) were considered superior, particularly owing to their simplicity, but as No. 3 was easily heaved by frost and snow it was considered suitable for milder climates, and the choice rested on the angle bars.

The Fisher bridge joint has been tested quite extensively, and is found to be very stiff vertically, but weak laterally, and its various parts are rather expensive and more complicated than the angle bars. For these reasons it is not likely to find extensive favor. The Churchill joint of N. & W. R. R. is probably the most efficient joint yet designed as far as stiffness, etc., and is intended for use with 60 ft. rails. Otherwise it would be too expensive and complicated for ordinary use. The other joints shown appear to have good points, but are of less tried merit. (Also see *Engineering News*, page 178, Vol. I., 1891, for Paterson rail joint.)

We may expect, ultimately, to obtain a joint as strong as the rail itself, but how simple it can be made is for the future to show.

ARTICLE 6.—RAIL FIXTURES, ETC.

The weak spot of our track is its attachment to the ties by ordinary track spikes. Their heads are often cracked by excessive driving, re-spiking is frequent, and the ties get split and rotten much sooner than they would naturally, and while Greer, Goldie, curved, interlocking and other special spikes are improvements on the dog spike, yet the final solution would seem to be in some positive fastening such as wood screws or fang bolts, such as are used to hold rail chairs to the ties on British roads, and while tie plates and selected oak ties are keeping off the evil day, yet as speeds get higher and engines heavier, demanding a high stiff rail, this must be done by heavy traffic roads sooner or later, either with woodén ties and tie-plates, or with steel ties and bolts.

Tie plates (such as Goldie, Servis, Standard, Sandberg, etc.) will enable roads even with heavy traffic to use soft wood ties and a high stiff rail with narrow base (see N. Y. C. & H. R. R. R. section), and will prolong the life of ties. They are being adopted rapidly, some roads using them on curves only, others for the whole track. Wood screws for holding track are of steel, seven inches long, with thread for $4\frac{3}{4}$ inches, $\frac{3}{4}$ inch diameter, and have a pulling resistance of about six tons. Fang bolts are attached by boring holes through the ties, and screwing the bolts, which have heads on them suitable for holding down a rail, into a nut, with a fang on it. This fang grips into the wood on the under side of the tie, which prevents it turning or loosening.

The vibration caused by passing trains would soon loosen the ordinary nut on the bolts which fasten the angle bar joint to the rails, and, in order to prevent this, many devices have been tried. The double nut is not effective. A gravity lock outside the ordinary nut in the form of an eccentric nut is much better, and Young's patent has been used quite extensively, but the spring nut lock, which consists of one turn or a little more, of a strong steel spiral, with two cutting lips taking hold of angle bar and nut as the nut is screwed on, on top of the nut-lock, is the kind generally used, and being simple, cheap and effective, is likely to remain the favorite kind in use.

ARTICLE 7.—SWITCHES AND FROGS.

Outlines of various designs for passing a train from one track to another are given on plate XXVII., but of course there are various forms of attachments differing in detail only.

(1) The Stub switch consists of two movable rails, A B, with the ends B supported, and free to slide on

plates for a lateral distance of five inches, called "throw." These switch rails or points are from 10 to 25 feet long, depending on the frog distance, B C, and the angle of the frog C. The guard rails, D D, prevent derailment at the throat of the frog. The stub switch works for a three-throw as easily as for a two-throw turnout, and can be made into a safety switch (see Cook Switch, Plate XXVII., and Dunn Switch, *Engineering News*, Vol. II., 1890, page 174), and is considered to be more durable and easily kept in working order with snow and ice than are the many forms of split rail switches.

(2) Dooley's stub switch is a modification which makes easier riding by having one point longer than the other, substituting two jolts for one severer one, but it is not as rigid as the ordinary stub switch.

(3) Nicoll half-safety switch is a compromise between the stub switch and a split or Lorenz switch. It is not at all a strong or secure switch, as the two rails are not opposite each other. Its advantages are not very obvious

(4) Lorenz safety switch is the model of various split switches. Both rails are feathered down so as to fit close up against solid rails. One is a main line rail, the other for the siding, connected so as to act together. This switch is adapted to position where the traffic is considerable on the branch line, or turnout, and in climates not troubled with ice or snow, but the split rails or points wear out rapidly, and it is more complicated when applied to three-throw turnouts, necessitating two sets of switch rails, stands, etc., set one ahead of the other, in which case neither of the main line rails are solid. The Stewart switch (*Engineering News*, Vol. I., 1895, page 59) has a special feature in making the switch rails by bending over solid-headed rails, instead of planing them down to a point. It is claimed this will give durability and rigidity.

(5) Ainsworth safety switch is made by giving the solid siding rail a sharp bend or recess, and the corresponding switch rail is left square ended, thus providing a more solid track for the main line, and a more durable switch rail. This form is adapted to branch lines having little traffic.

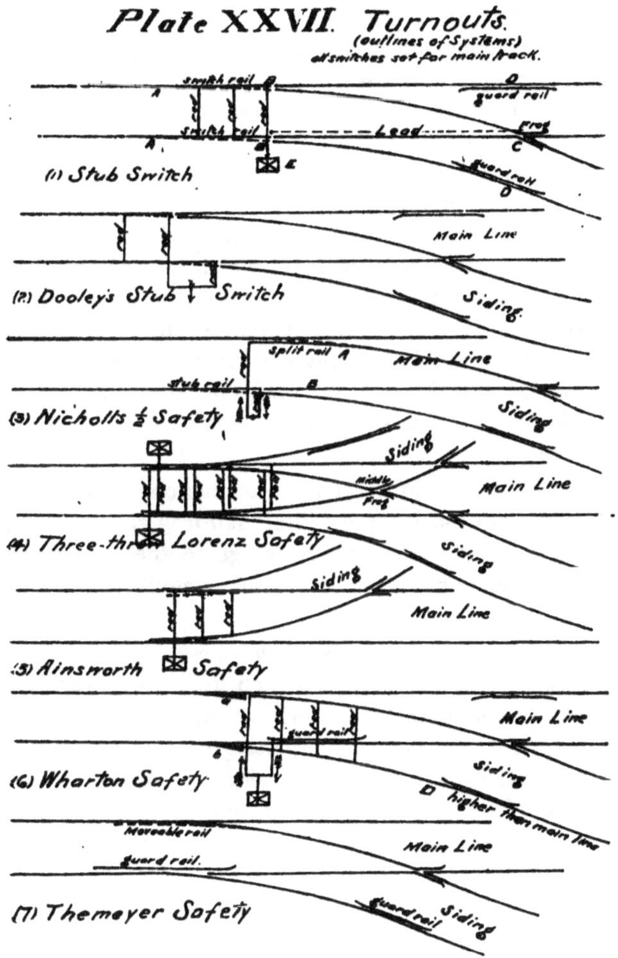
Plate XXVII. Turnouts.
(outlines of Systems)
all switches set for main track.
switch rail
guard rail
Lead
Frog
guard rail
(1) Stub Switch
Main Line
(2) Dooley's Stub Switch
Siding.
split rail A
Main Line
stub rail
B
(3) Nicholls ½ Safety
Siding
Siding
middle Frog
Main Line
(4) Three-throw Lorenz Safety
Siding
Siding
Main Line
(5) Ainsworth Safety
Main Line
guard rail
Siding
(6) Wharton Safety
D higher than main line
Moveable rail
Main Line
guard rail.
guard rail
Siding
(7) Themeyer Safety

(6) Wharton safety switch is used for heavy main line traffic. It gives a solid main track. The siding rails lead the wheels onto blocks (a.b.) higher than the main line rails, and fall down on to the main line, while in facing the switch the wheels are first lifted by the blocks (a.b.) and then carried over the main line rails by the wheel tread riding on the high rail D.

The Macpherson switch (Plate XXVIII.) is a modified Wharton coming into use on the Can. Pac. Ry. The main line is solid, and the train is thrown onto the siding by having the outside movable rail higher than the main line, and a movable guard rail which is also higher than the main line, but which is thrown into position only when the switch is set for the siding. This design also includes a special form of frog, which is a sliding plate brought into position by means of bell-crank levers and rods operated from the switch stand, when set for siding; when set for main line the plate is clear of the main line, leaving the main line solid at this point also. This design has been in use since 1892, and it has proven itself very satisfactory and durable.

(7) The Themeyer safety switch has one movable split rail, and a stationary split rail or half-frog and guard rail. The movable rail and guard rail guide the wheels onto the siding when set for it. It is successfully used on the B. & O. R. R.

The main object of safety switches is to make it safe for a train to trail through a switch from the siding, when it is set for the main line, or vice versa, and this is accomplished, with split switches, by using springs which allow the movable rails to be forced aside just enough to pass the wheel flanges through. The springs then force the switch points back to the position for which the stand and signal are set.

Other special switches of tried merit are the cam automatic, in which the split rails are fixed, and the solid ones move horizontally (see *Eng. News*, vol. I., 1890, page 489), and the Duggan switch, which has two knuckle-jointed vertical moving split rails. (See *Eng. News*, vol. I., 1893, page 390.)

Frogs.—Formerly cast steel solid frogs were common, but as they were more liable to crack, and when worn in one part were unfit for use, they were soon supplanted by frogs made up of pieces of steel rail fitted and bolted together onto a flat steel base plate—any worn part can be easily replaced. Such solid or stiff frogs are in most general use, but on main lines having heavy traffic, those turnouts with light traffic are now generally fitted with spring frogs (see Plate XXVIII.) in which either the "point" or the guard rail are movable, and the main line is normally a solid track. A train to or from the siding forces the frog open momentarily, and a spring brings it back again as soon as the train has passed, leaving the main line again solid. The defect in many of these spring frogs is the tendency to derail wheels with worn treads and flanges, by forcing open the spring frog when a train is on the main line. It is claimed that the Vaughan spring frog, used on the Penn. R. R., overcomes this difficulty by blocking up the tread. Other spring frogs of special features of merit are the Monarch, Ramapo and Pegram, described in the *Engineering News* since 1890.

Turnout Calculations.—The "lead" is the distance KD (Plate XXVIII.) from the switch stand to the frog point. The fixed end of switch rails is the "heel," and the movable one the "toe." The "throw" is the amount which the switch stand rod moves the "toe" of the switch. It is 5 inches for stubs and 3 inches for split switches. To designate a frog angle $E\ D\ F$ the ratio $\frac{ED}{EF}$ is called the frog number (i.e.) if $\frac{ED}{EF}=6$, then the frog is called a No. 6, the ordinary numbers in use are 8, 9, and 10 for main lines, and 5, 6 and 7 for crowded yards and sidings. The middle frog is a special one, derived from the others by calculation or from a large-scaled plan:

(1) To calculate the lead from the frog number we have (see Wicksteed, Trans. C. Soc. C. E.) $D\ M$ (Fig. 1.) $=\frac{\text{gauge}}{\sin \alpha}$ or approximately $A\ D = \frac{2 \times \text{gauge}}{\sin \alpha}$ but, for small angles $\frac{1}{\sin \alpha}$ = frog number = N and gauge = g = 4·75 ft., approximately, then frog distance $= 2.g, N = 9{\cdot}5\ N$ – – – – (A).

(2) *To find the length of the movable rails.*—Offsets to a circle from a tangent vary as the square of the distance from tangent point, and taking gauge as 57 inches and throw as 5 inches we have

$$\frac{\text{slide rails}}{\text{frog distance}} = \sqrt{\frac{5}{57}} = \frac{3}{10} \text{ approx. } (B)$$

also, for stub switch, lead $= \frac{7}{10}$ frog distance (*C*) which equations give all necessary data for a simple turnout for a stub switch. The frog for a very short distance is straight, and the slide rail is often practically straight, but by using a long rail and spiking the fixed portion, the movable part will bend to a curve.

If split switches are used Fig. 2 will apply, and the movable rail being, necessarily, straight, is from *B* to *C* only, is tangent to the circle at *B*, and is half as long as a stub switch movable rail, also in this case the switch stand is at a different place *K. C.* and we have

Switch rail $= \frac{1}{2} \times \frac{3}{10} = \frac{3}{20}$ frog distance (*D*).
Lead = frog distance $- A\ C = \frac{17}{20}$ frog distance (*E*).

EXAMPLES.

(*a*) Stub switch No. 8 frog—
Frog distance $= 8 \times 9{\cdot}5 = 76$ feet by (*A*).
Slide rail $= \frac{3}{10} \times 76 = 22\frac{8}{10}$ feet by (*B*).
Lead $= 76 - 22{\cdot}8 = 53\frac{2}{10}$ feet by (*C*).

(*b*) Split switch No. 9 frog—
Frog distance $= 9 \times 9{\cdot}5 = 85.5$ feet by (*A*).
Lead $= \frac{17}{20} \times 85.5 = 72.6$ feet by (*E*).
Slide rail $= \frac{3}{20} \times 85.5 = 12.9$ feet by (*D*).

NOTE.—These distances can be varied by a small percentage without affecting the running of the trains.

(3) Middle frog calculations, Fig. 3, Plate XXVIII. First let the two turnouts *AE AF* be of same degree of curve and start from same switch stand, then $\frac{AD}{AC} = \sqrt{\frac{1}{2}}$ or $A\ D = \frac{71}{100}$ *A C*, which gives us the middle frog distance from the frog-distance *AC*, which equation (*A*) determines.

Plate XXVIII.

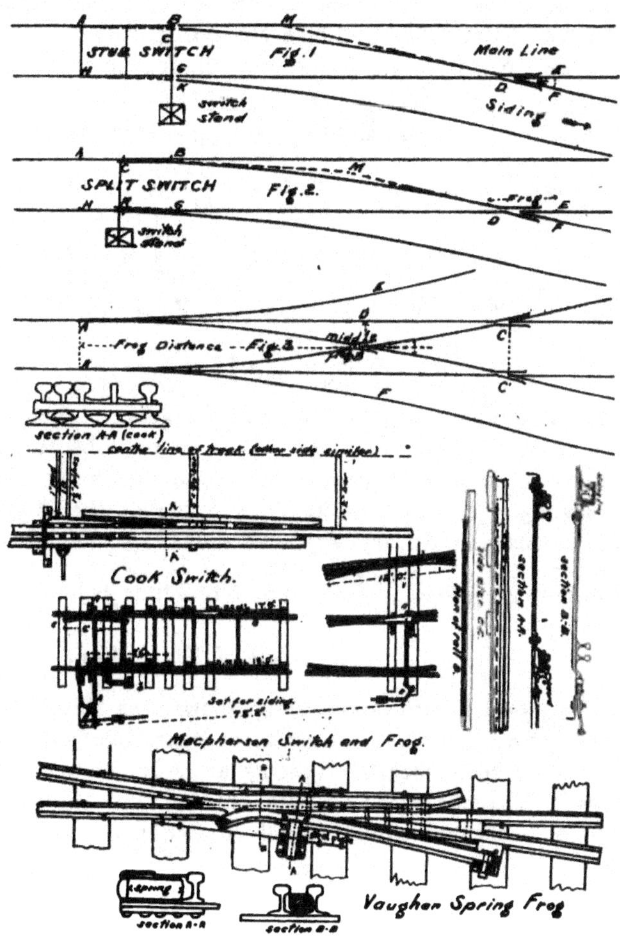

Also for small angles the angle of the middle frog will be $2 \times \frac{71}{100}$ frog angle at C = 1.42 frog angle C, and the number of the frog will be $\frac{1}{1.42} \times$ frog number C = ·703 frog number C.

Second let the turnouts be of different sharpness, and let one begin say 6 feet ahead of the other, let the right hand turnout start first and be a No. 8 frog, and the left hand one a No. 10 frog. Call the middle frog distance x.

The two turnout frog distances are

$8 \times 9{\cdot}5 = 76$ feet, and $6 + 10 \times 9{\cdot}5 = 101$ feet.

Offset from one tangent to $B = \left(\frac{x}{76}\right)^2 \times 4{\cdot}75$ ft.

" " other tangent to $B = \left(\frac{x-6}{95}\right)^2 \times 4{\cdot}75$ ft.

But both offsets added = gauge = 4.75 ft., therefore

$$\left\{\left(\frac{x}{76}\right)^2 + \left(\frac{x-6}{95}\right)^2\right\} 4{\cdot}75 = 4{\cdot}75$$

and solving this quadratic equation we get $x = 61{\cdot}6$ ft. Also we can determine its lateral position by substituting in either of the above equations this value of x, in this case these are 3.13 ft. and 1.62 ft. The angle of the middle frog, in this case, can be calculated thus:

$$\text{Middle frog angle} = \frac{61.6}{76} \text{ angle of No. 8 frog} + \frac{55.6}{95} \text{ angle of No. 10 frog.}$$

In crowded yards and with split switches these conditions prevail, and many much more intricate calculations ore often needed when the turnouts are from curves, and cross other tracks which are also curving, but these can often be best obtained by carefully drawn plans to large scale.

Mileage	§ 16,687	
Population	‖ 5,300, 00	318
Square miles	‖ 1,609, 00	
Population per square mile	3.3	
Capitalization—		
Bonds	$348,834,000	$20,904
Preferred stock	107,235,000	6, [illegible]6
[illegible]n stock	260,376,000	15,603
Loans and floating [illegible]ts	37,068,000	2,221
Bonus and [illegible]ent aid	173,329,000	10,386
Total	$926,842,000	5[illegible]40
Earnings, etc.—		
Passenger	$13,929,350	$ 835
Freight	[illegible] 3[illegible]0	2,009
Other	4,901,830	294
Total	$52, [illegible]80	$3,138
Working expenses	35,168,660	2,108
Net earnings	$17,184,620	$1,030
Working expenses—		
Per cent. gross earnings	67	
Net earnings—		
Per cent. gross capitalization	1.85	
Fixed charges—		
Bond interest	* $16,014,370	
Other interest on loans, etc	† 1,853,400	
Total	$17,867,770	
Apparent net income	− 683,150	
Passenger charge per mile	‡ 2⅛c	
Freight charge per ton mile	‡ 1⅛c	
Earnings per passenger train mile	80c	
Earnings per freight train mile	$1.42	

Average cost per train mile	77c	
Gross [illegible]ngs per inhabitant per year	$9.88	
Passenger traffic—		
Number of passengers	16,171,340	969
Passenger cars	1,773	$\frac{11}{100}$
Average [illegible]ul (miles)	+ 34	
Average load	+ 32	
Passenger cars per 1,000, 00 [illegible]ssengers	109	
Passenger train miles	17,237,970	
Passenger [illegible]ns per day ([illegible]h way)	× $1\frac{2}{5}$	
Freight traffic—		
Tons of freight	25,300,330	1,516
[illegible]ght cars	5754	
Average haul (miles)	+ [illegible]6	
Average load [illegible]	+ 114	
Freight 00 tons	2,280	
Freight	23,595,000	
Freight [illegible]ns per day (each way)	× $1\frac{95}{100}$	
[illegible]d traffic—		
Various kinds of cars	90	
[illegible]d train [illegible]ge	9[illegible]7,877	
[illegible]d trains per day (each way)	× $\frac{45}{100}$	
[illegible]l trains per day (each way)	× $3\frac{4}{5}$	
Engines—		
[illegible]l number	2,096	
[illegible] [illegible]early engine mileage	÷ ,2[illegible]0	
Employees	‡ 57, 00	
" killed	76	
" injured	579	
Passengers killed	7	
" injured	67	
" killed, 1 in	2,310,200	
" injured, 1 in	241,000	

§ Not including extra tracks or sidings.
‖ Only explored regions of Canada counted.
* Includes 4½% interest on $15,000,000 bonds on which rate of interest is not given.
† Estimated at 5%.
‡ Estimated.
+ Based on assumed average passenger and freight rates.
× 365 days per year.
÷ Including switching engines.

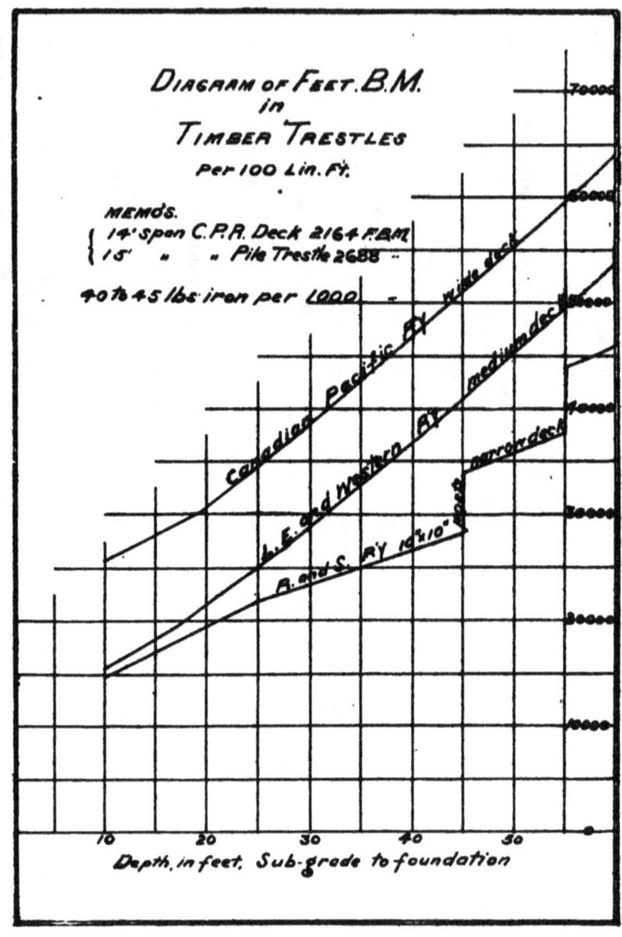
DIAGRAM OF FEET. B.M.
in
TIMBER TRESTLES
Per 100 Lin. Ft.
MEMO'S.
14' span C.P.R. Deck 2164 F.B.M.
15' " " Pile Trestle 2688 "
40 to 45 lbs iron per 1000
Canadian Pacific R'y wide deck
L. E. and Western R'y medium deck
R. and S. R'y 10"x10" narrow deck
70000
60000
50000
40000
30000
20000
10000
0
10
20
30
40
50
Depth, in feet, Sub-grade to foundation

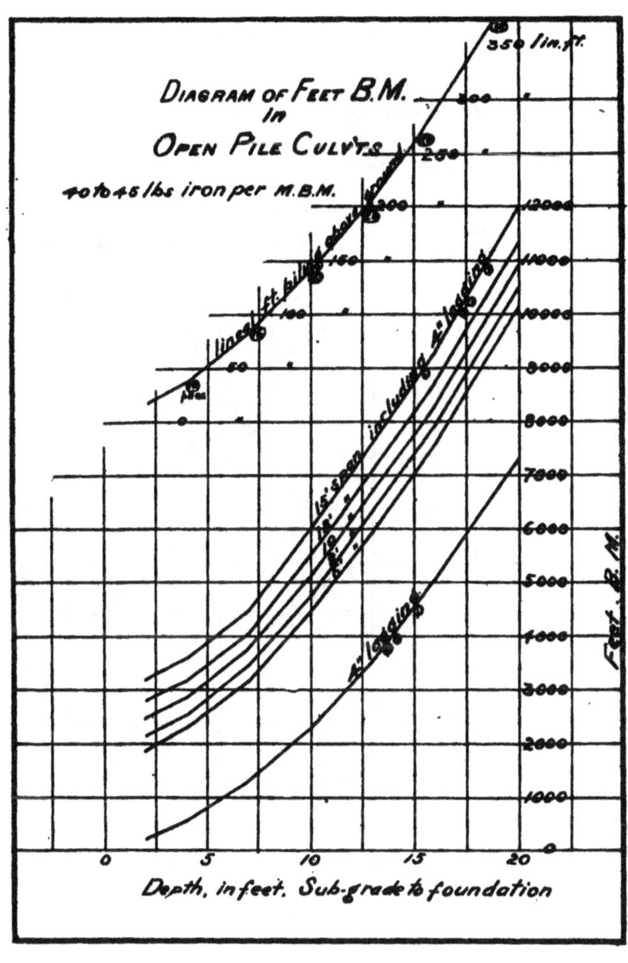
DIAGRAM OF FEET B.M.
in
OPEN PILE CULV'TS
40 to 45 lbs iron per M.B.M.
350 lin. ft.
300
250
200
150
100
50
0
12000
11000
10000
9000
8000
7000
6000
5000
4000
3000
2000
1000
0
0
5
10
15
20
Depth, in feet. Sub-grade to foundation
Feet B. M.
4" lagging
15' span including 4" lagging

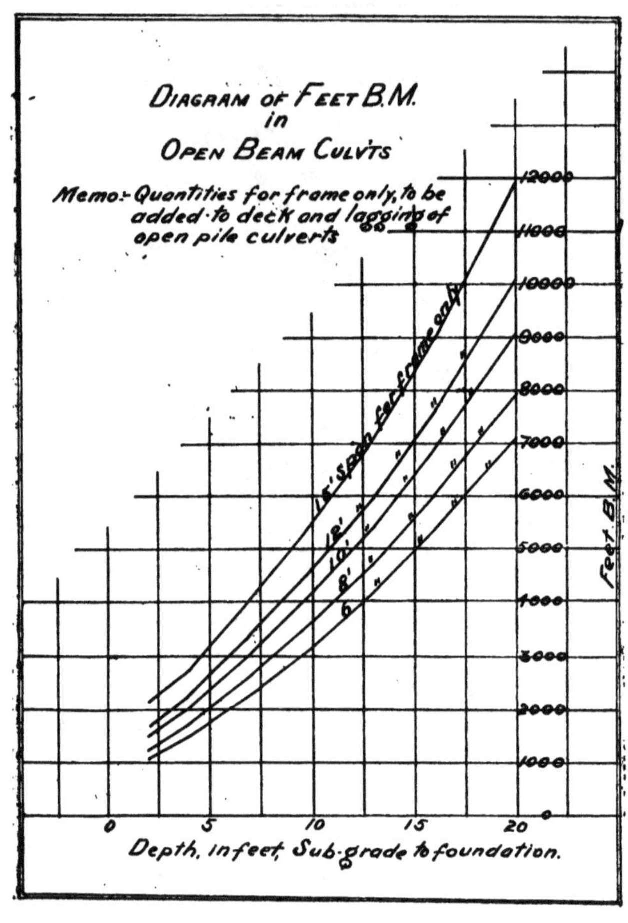
Diagram of Feet B.M.
in
Open Beam Culvts
Memo:- Quantities for frame only, to be added to deck and lagging of open pile culverts
15' span feet frame only
12' "
10' "
8' "
6 "
12000
11000
10000
9000
8000
7000
6000
5000
4000
3000
2000
1000
0
Feet B.M.
0
5
10
15
20
Depth, in feet, Sub-grade to foundation.

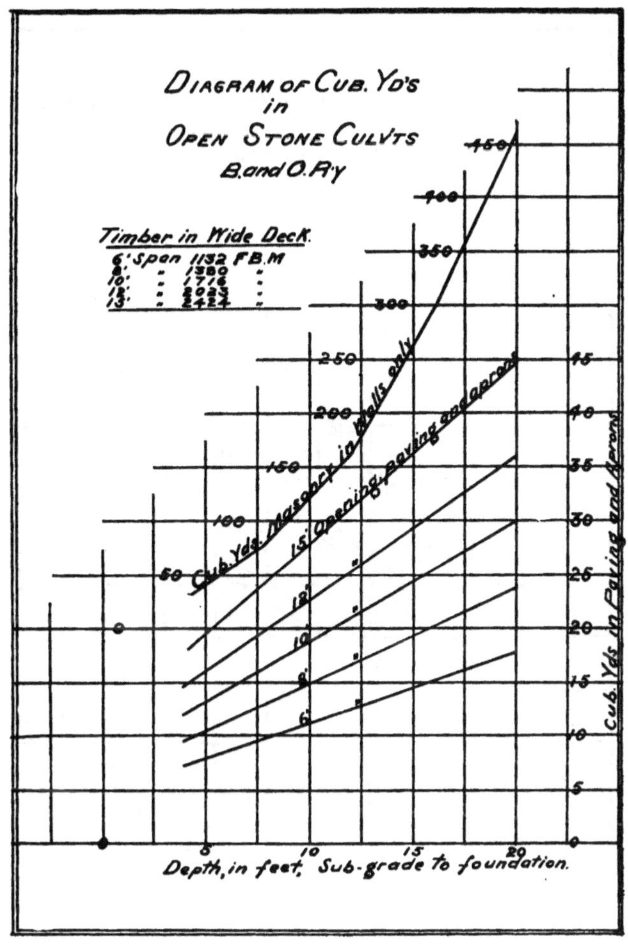
DIAGRAM OF CUB. YD'S
in
OPEN STONE CULVTS
B. and O. R'y
Timber in Wide Deck.
6' Span 1132 F.B.M.
8' " 1380 "
10' " 1716 "
12' " 2023 "
13' " 2424 "
Cub. Yds. Masonry in Walls only
15' Opening, paving and aprons
12'
10'
8'
6'
50
100
150
200
250
300
350
400
450
0
5
10
15
20
25
30
35
40
45
Cub. Yds in Paving and Aprons
Depth, in feet. Sub-grade to foundation.

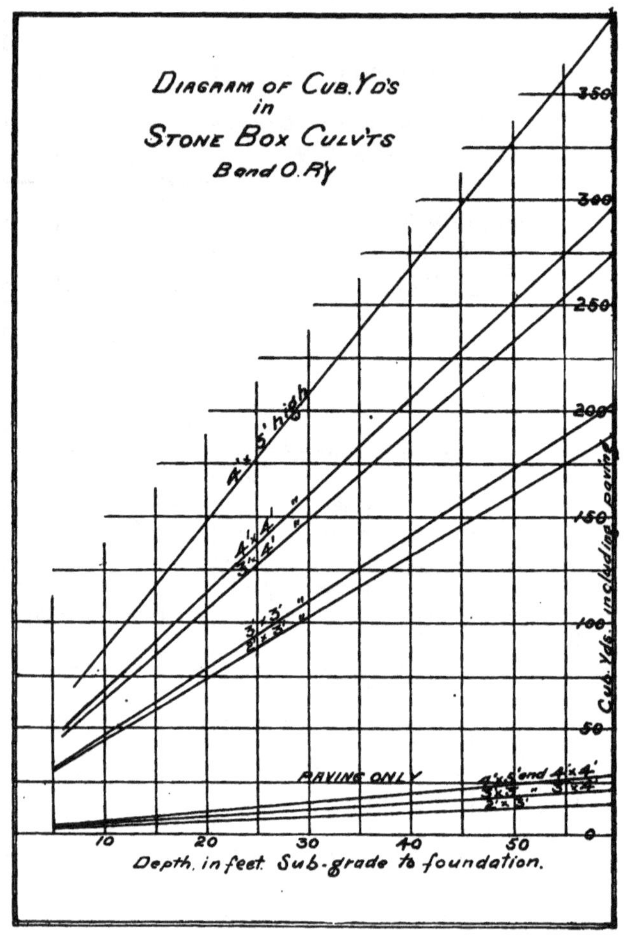
DIAGRAM OF CUB. YD'S
in
STONE BOX CULV'TS
B and O. RY
4'x5' high
4'x4' "
3'x4' "
3'x3' "
2'x3' "
PAVING ONLY
4'x5' and 4'x4'
3'x3' " 3'x4'
2'x3'
350
300
250
200
150
100
50
0
10
20
30
40
50
Cub. Yds. including paving
Depth, in feet Sub-grade to foundation.

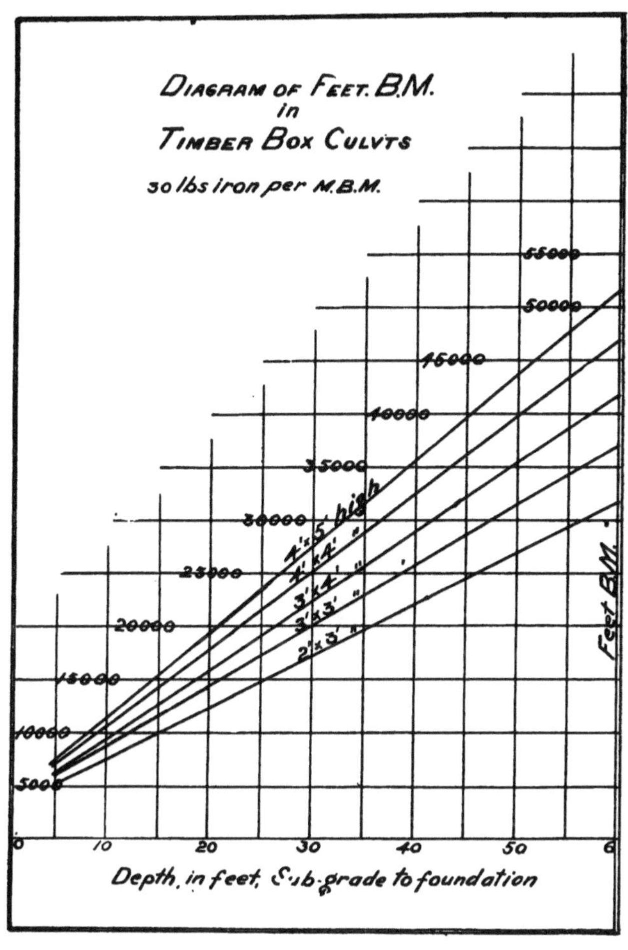
DIAGRAM OF FEET. B.M.
in
TIMBER BOX CULVTS
30 lbs iron per M.B.M.
55000
50000
45000
40000
35000
30000
25000
20000
15000
10000
5000
4'x5' high
4'x4' "
3'x4' "
3'x3' "
2'x3' "
Feet B.M.
0
10
20
30
40
50
6
Depth, in feet, Sub-grade to foundation

INDEX.

www.ingramcontent.com/pod-product-compliance
Lightning Source LLC
LaVergne TN
LVHW021940220826
846092LV00010B/1191